Healing *touch*

Healing *touch*

A complete guide to
the use of touch therapies
that promote well-being

MARCUS & MARIA WEBB

 A GODSFIELD BOOK

Library of Congress Cataloging-in-Publication Data Available

10 9 8 7 6 5 4 3 2 1

Published in 1999 by Sterling Publishing Company, Inc.
387 Park Avenue South, New York, N.Y. 10016
© 1999 Godsfield Press
Text © 1999 Marcus and Maria Webb

Marcus and Maria Webb assert the moral right to be identified
as the authors of this work.

Distributed in Canada by Sterling Publishing
c/o Canadian Manda Group, One Atlantic Avenue, Suite 105
Toronto, Ontario, Canada M6K 3E7
Distributed in Australia by Capricorn Link (Australia) Pty Ltd
P O. Box 6651, Baulkham Hills, Business Centre, NSW 2153,
Australia

Printed and bound in Hong Kong

Sterling ISBN 0-8069-9917–9

Contents

Introduction

Our highly developed sense of touch is fundamental to a healthy and balanced lifestyle. From the very first minutes of life our nervous system starts to detect and store information about the surrounding environment; its taste, texture, smell, sound, and color. This accumulation of sensory experience grows and forms the basis of our experience. In time we learn to make associations between texture and taste, smell and color, and other combinations until we are able to form a complete mental database of learned sensations. From this we know instinctively how to respond to the world around us; however hard we try, it is impossible to detach ourselves from the basic principle of sensory integration.

Health – from the holistic perspective – demands that all aspects of the individual are functioning effectively; spirit and body have to be in harmony to maximize an individual's well-being. Modern medicine offers life-saving procedures but it can fail to

appreciate the importance of simple touch in its approach to healthcare. The transforming power of touch can calm a crying baby in just a couple of seconds as the mother holds the infant in her arms, offering comfort and love. Similarly, when a stressed individual receives a massage or an aromatherapy treatment, their entire body relaxes and emotional and physical tensions are released. The profound sense of relief and tranquility experienced after a massage demonstrates just how much wasted energy was trapped in the muscles. For this reason the subject may find the treatment very draining.

In writing this book we wanted to provide a complete overview of the various hands-on treatments available, reviewing their history, philosophy, and applications. From these options we selected four systems of treatment that could be performed safely and effectively in the home and have suggested step-by-step therapies that you can perform on yourself, your partner, or friends and family. With its comprehensive lists of indications, safety precautions, and an extensive resource directory we are confident that this book will become an indispensable manual promoting a lifetime of physical and mental well-being.

Marcus Webb BSc DO

Osteopath

Maria Webb LCSP

Physical Therapist

How to use this book

Touch is a biological necessity as well as a physical pleasure, but most of us do not realize its importance. This book will appeal to anyone interested in the various touch therapies including massage, aromatherapy massage, Chinese massage, acupressure, reflexology, shiatsu, Hellerwork, Rolfing, reiki, healing, and kinesiology. Divided into five comprehensive sections, the book covers all aspects of healing touch and its benefits.

Section 1, "The biology of touch," explores the basic need for human contact and the benefits to our health and well-being that it brings. Whether you wish to relieve muscle tension, reduce stress, or give a boosting facial massage, you'll find easy step-by-step sequences to follow.

The history of each therapy is covered, as are the principles and practicalities behind the treatment. The section entitled "Touch therapy for well-being" gives you practical advice on setting the scene, preparing yourself, and using specific therapies for relaxation, vitality, and sensuality.

You are then taken through the stages of life – from birth to old age – and advised which therapies are beneficial for conditions specific to your age. The last section, "Touch therapy for body areas," covers those therapies you can practice at home – massage, aromatherapy massage, acupressure, and reflexology. Contraindications tells you what to avoid, after which you will find useful addresses and a reading list for further information.

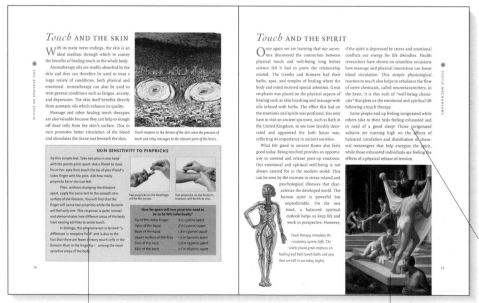

"The biology of touch" looks at the physiological and spiritual effects of touch. Boxes contain step-by-step practical exercises or advice.

Clear text explains the importance of touch to human beings

Boxed practical information for useful exercises

Historical illustrations help explain the background to therapies

Concise text explains the principles of the therapy

Box pulls out key information on the benefits of the therapy

Essential reference map for the therapy

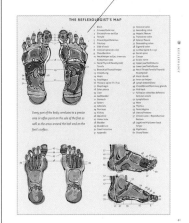

This section covers all the main touch therapies, giving a thorough history of each. It also explains the principles behind them and how practitioners work, and shows practical information, such as the basic massage strokes. It also reveals the particular benefits of each therapy.

"Touch therapies for life" takes you through the generations and reveals the aspects of touch therapy that will help throughout. The emphasis is on therapies you can practice at home.

Acupressure illustration shows position of key acupoint

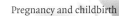

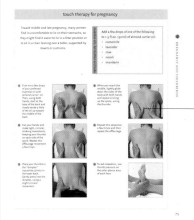

Step-by-step photographic sequence explains a specific touch therapy treatment

The last section covers all the major body areas from the immune system to digestion. It explains the importance of the balanced and efficient working of each area to ensure harmony throughout your body. Specific photographic step-by-step treatments are given for key conditions.

Boxed information on useful essential oil combinations

Step-by-step photographic sequence explains a specific touch therapy treatment

Key reflexology points highlighted

1

THE BIOLOGY OF TOUCH

We all have the age-old instinct to rub a painful area, and the immediate sensation produced by this action gives us a feeling of comfort and well-being. However, modern research has revealed aspects of touch and touch-based therapies that give validity and new insights into this ancient wisdom and practice.

In this section, we look at the science and biology of touch. For such a fundamental method of healing, touch has only recently received the interest of the scientific community. As the studies progress, more and more evidence is accumulating in favor of touch for the promotion of total good health.

It is interesting to note that beneficial effects can be felt by the giver as well as the recipient of a touch-based therapy. This aspect is often overlooked, with the focus of interest being on the recipient rather than the giver. Later in the book, we offer suggestions on how to maximize the effectiveness of your therapeutic approach by training yourself to become aware of your own body, with its strengths and its weaknesses.

Like many aspects of life, knowledge is the key to understanding. By reading this section carefully, you will gain greater insight and be able to benefit from the later chapters of the book. In time, and with practice, you will be able to formulate your own personal approach to therapeutic touch using these simple principles as your foundations.

Touch mechanisms

1 Humans are by nature intimately related to each other and their environment by a sense of touch. The ability to feel and discover the world around us can give the blind or visually impaired "eyes" in the fingers, and it is often said that blind massage practitioners make the best therapists. Touch is such a vital sense that vast areas of the brain are solely devoted to the integration of touch messages with other brain areas associated with emotions, memories, and imagination.

THE DESIRE TO LEARN

Of all the senses, touch must be one of the most fundamental to our well-being. We tend to reach out unconsciously to touch new things. This instinctive desire to learn with our hands about texture, size, form, and temperature can be traced back to our earliest years as newborn babies when we continuously touch and feel our way around our new environment with our hands and mouths.

Building up a mental store of data in response to our environment is among the first stages of brain development. Touch, therefore, forms a central point about which the individual develops a feeling for his or her place in the world; touch is an integral aspect of a person's general sense of well-being and security.

Touch is one of the first senses a baby has, and a child's first introduction to the world presents many different new objects and textures to touch. This sense offers vital contact between child and parent and is crucial to developing relationships as well as learning.

Touch AND THE BRAIN

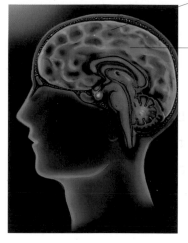

Cerebral cortex

Touch center of the brain

Recent technological developments have given us further information about the brain and our sensory skills.

All you need to do to discover how a minute movement is magnified into a perceivable sense by the brain is touch a single tiny hair on the back of your hand. The movement is so small that the skin surrounding the hair cannot be seen to move, but the sensation is still easily felt. Skin that is devoid of hair is still highly sensitive, having specialized touch cells just beneath its surface.

When we consider the sense of touch, it must be appreciated that what we feel as touch may come from many different stimuli. For example, the mechanical effects of pressure applied to the skin are detected by different nerves from those that detect a vibration. Nerve endings in the skin, therefore, have evolved in a variety of ways to detect certain types of touch stimuli. In a similar way, the nerves' pathway to the brain is quite different for the sensation of pressure compared to that of vibration. Our bodies have evolved a highly complex filtering system that diverts certain touch sensations to either fast or slow pathways to the brain. This is probably a Darwinian evolutionary adaptation, which is dependent on how important each feeling is to our survival. For example, the perception of heat can provide a warning that we are about to burn ourselves, which, in turn, could cause tissue damage. Such a sensation is not always processed by the brain but rather as a reflex action in the spinal cord. The action of dropping the hot object, or pulling quickly away from the flame, can be performed in a split second. If we had to stop and think carefully about it and make connections with other brain centers, the time delay, albeit measured in microseconds, could be long enough for the anticipated damage to occur.

Other touch sensations, however, require quite a lot of "higher" processing. The brain needs to use its ability to integrate "learned" sensations with stored memories, as in the case of a blind person "reading" Braille, for example.

The complex system of raised-point writing to be read by the visually impaired was developed by Louis Braille in 1829. He himself had been blind since the age of three.

Touch AND THE SKIN

With its many nerve endings, the skin is an ideal medium through which to convey the benefits of healing touch to the whole body.

Aromatherapy oils are readily absorbed by the skin and they can therefore be used to treat a large variety of conditions, both physical and emotional. Aromatherapy can also be used to treat general conditions such as fatigue, anxiety, and depression. The skin itself benefits directly from aromatic oils which enhance its quality.

Massage and other healing touch therapies are also valuable because they can help to slough off dead cells from the skin's surface. This in turn promotes better circulation of the blood and stimulates the tissue just beneath the skin.

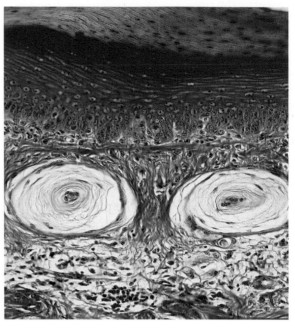

Touch receptors in the dermis of the skin sense the pressure of touch and relay messages to the relevant parts of the brain.

SKIN SENSITIVITY TO PINPRICKS

Try this simple test. Take two pins in one hand with the points 5mm apart. Ask a friend to close his or her eyes then touch the tip of your friend's index finger with the pins. Ask how many pinpricks he or she can feel.

Then, without changing the distance apart, apply the same test to the smooth skin surface of the forearm. You will find that the finger will sense two pinpricks while the forearm will feel only one. This response is quite normal and demonstrates how different areas of the body have varying abilities to sense touch.

In biology, this phenomenon is termed "a difference in receptive field" and is due to the fact that there are fewer sensory touch cells in the forearm than in the fingertip – among the most sensitive areas of the body.

Two pinpricks on the forefinger will be felt as two.

Two pinpricks on the forearm, however, will be felt as one.

How far apart will two pinpricks need to be to be felt individually?

Tip of the index finger	⅛ in (5mm) apart
Palm of the hand	⅓ in (10mm) apart
Back of the hand	1 ⅙ in (30mm) apart
Upper surface of the foot	1 ⅗ in (40mm) apart
Skin of the neck	2 ⅛ in (55mm) apart
Skin of the back	2 ½ in (65mm) apart

Touch AND THE SPIRIT

Once again we are learning that our ancestors discovered the connection between physical touch and well-being long before science felt it had to prove the relationship existed. The Greeks and Romans had their baths, spas, and temples of healing where the body and mind received special attention. Great emphasis was placed on the physical aspects of healing such as skin brushing and massage with oils infused with herbs. The effect this had on the emotions and spirits was profound. You only have to visit an ancient spa town, such as Bath in the United Kingdom, to see how lavishly decorated and appointed the bath house was, reflecting its importance in ancient societies.

What felt good in ancient Rome also feels good today. Being touched provides an opportunity to unwind and release pent-up emotions. Our emotional and spiritual well-being is not always catered for in the modern world. This can be seen by the increase in stress-related and psychological illnesses that characterize the developed world. The human spirit is powerful but unpredictable. On the one hand, a balanced spiritual outlook helps us keep life and work in perspective. However,

Touch therapy stimulates the circulatory system (left). The Greeks placed great emphasis on healing and built lavish baths and spas that are still in use today (right).

if the spirit is depressed by stress and emotional conflicts our energy for life dwindles. Health researchers have shown on countless occasions how massage and physical interaction can boost blood circulation. This simple physiological reaction to touch also helps to rebalance the flow of nerve chemicals, called neurotransmitters, in the brain. It is this rush of "well-being chemicals" that gives us the emotional and spiritual lift following a touch therapy.

Some people end up feeling invigorated while others take to their beds feeling exhausted and in need of a good sleep! Those invigorated subjects are running high on the effects of balanced circulation and distribution of chemical messengers that help energize the spirit, while those exhausted individuals are feeling the effects of a physical release of tension.

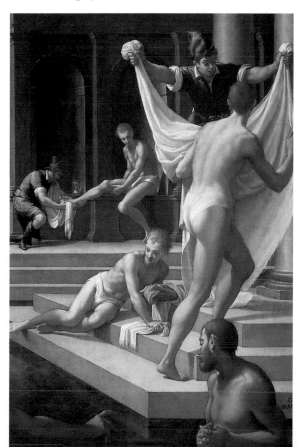

15

The healing effects: *giving*

Much is known about the benefits gained by those receiving a touch-based therapy, but little attention has been paid to the effects of giving gained by the therapist. Some claim that the physical effort involved in giving such an intimate treatment can only deplete the giver's energy reserves, while others suggest that the act of giving actually energizes the body. As long as you are focused and balanced beforehand, you can increase your sense of well-being by giving a touch therapy.

The concept of energy depletion is an important consideration when you are planning treatment. Interaction on a physical, as well as a mental, level makes demands of your body. If you are not physically fit or are feeling tired, you may find the task exhausting and will not be able to give an effective treatment.

Body energy can fluctuate from day to day. In women, body energy is very sensitive to the hormonal changes associated with the menstrual cycle. Being low in energy makes you vulnerable to becoming drained after giving treatment to a subject who may draw upon what little strength you have in reserve.

The hands can convey a lot of information about the person to whom they belong. Tensions in the body can manifest as tension in the hands, resulting in poor sensitivity and an inability to transfer healing energy. Being able to identify your character and personality type can help you give someone better touch therapy. Personalities are grouped into two main types, Type A and Type B (see the boxes opposite).

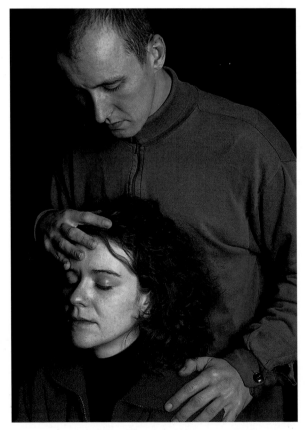

The potential benefit of giving touch therapy is often undervalued. The act of giving massage and healing can be as energizing as receiving touch.

PERSONALITY TYPES

TYPE A: If four or more of the personality traits below apply to you, it is likely that you fall into the energetic, perfectionist Type A group:

• easily irritated over trivial matters
• always doing things in a hurry
• critical of others
• angry with yourself over mistakes
• very competitive
• a tendency to interrupt and to try to finish sentences for others.

Certain aspects of your character may require attention in order to allow the true healing qualities of touch to come through and be of full benefit.

TYPE B: If fewer than four of the traits apply to you, you probably are a Type B personality. This character type is generally more suited to touch therapy, having a calmer and more relaxed attitude to life. They tend to enjoy caring for others and listening to their concerns.

Type A personalities are often easily irritated over trivial matters and become angry with themselves.

BALANCING EXERCISES

It is possible, however, to train yourself to become more relaxed, even if you are a typically impatient Type A personality:

• learn to slow down
• take time to listen and appreciate the feelings of others
• think before you speak
• avoid getting angry over the small annoyances in life
• study the art of meditation and try yoga
• appreciate that games are for enjoying, not always for winning
• accept that people are not perfect, so learn to acknowledge that mistakes do happen
• plan your time more realistically and don't clutter up your diary with appointments.

Balancing your character type between Types A and B will have a positive effect on your beliefs, attitudes, and habits, and is bound to improve your ability to give sympathetic and effective touch therapy to others.

Learn to relax to become a better "giver."

STIMULATING THE FLOW OF INNER ENERGY

Looking more closely at the effect of positive body energy, you must consider the ability we all possess to focus our healing energy and to draw upon it when needed. This, again, will require practice but once mastered it should form part of your daily routine.

This skill benefits those people you come into physical contact with, but it also boosts your own inner feelings of well-being.

When you are giving the healing touch, it is not a matter of simply using your hands. Your energy must flow out through your shoulders and upper arms to impart maximum benefit to the recipient.

focusing healing energy

It is important for both you and the person you will be treating that you "center" yourself beforehand.

1. Find a comfortable sitting position, with your feet placed firmly on the ground. Keeping your spine erect, close your eyes and clasp your hands loosely in your lap.

2. Begin by slowly and rhythmically breathing in through your nose and out through your mouth. Try to make the exhaling breath twice as long as the inhaling breath. As you concentrate on this breathing exercise, feel your hands and feet warm up and start to tingle as your energy flow is stimulated.

3. To make this exercise stronger, visualize the energy flow as a red glow that passes down your arms and legs with each exhaling breath. As the red glow settles in your hands and feet at the end of each exhaling breath, feel the warmth generating and spreading up toward your body. Continue to follow the rhythmic breathing exercise for 10 minutes or until you feel adequately energized and ready to proceed to the next step – putting your energy to good work.

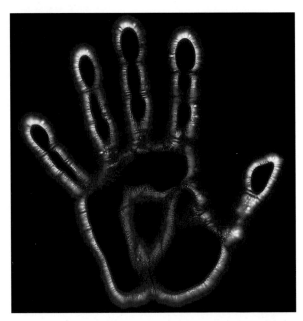

Kirlian photography is a method of producing prints of a person's aura: this reveals that person's emotional and physical disorders before any symptoms are seen. Kirlian photography is very useful to the touch therapist in that it highlights the areas of the body that need treatment.

PUTTING THE POSITIVE ENERGY TO WORK

Using the positive internal energy that was generated from the exercise on focusing healing energy, turn your attention to your physical state. Recruiting healthy muscle function and circulation lies at the very core of effective therapeutic touch.

Simple yoga exercises are an excellent way to achieve this. There are many forms of yoga to choose from, and all aim to achieve harmony between the body, mind, and spirit.

Correct breathing techniques are essential for getting the most from any yoga exercise because they encourage an improved circulation via efficient use of the lungs. To sample basic positions try the exercise known in yoga as the mountain posture (see below).

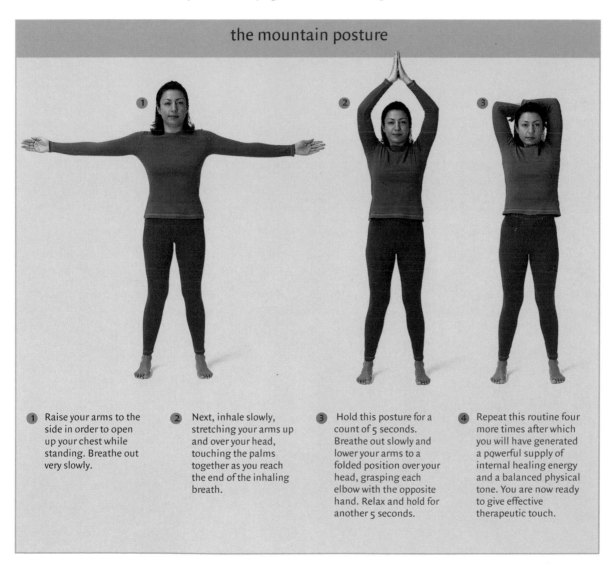

the mountain posture

1   Raise your arms to the side in order to open up your chest while standing. Breathe out very slowly.

2   Next, inhale slowly, stretching your arms up and over your head, touching the palms together as you reach the end of the inhaling breath.

3   Hold this posture for a count of 5 seconds. Breathe out slowly and lower your arms to a folded position over your head, grasping each elbow with the opposite hand. Relax and hold for another 5 seconds.

4   Repeat this routine four more times after which you will have generated a powerful supply of internal healing energy and a balanced physical tone. You are now ready to give effective therapeutic touch.

The healing effects: *receiving*

Our instinctive reaction to pain, both physical and emotional, is to reach out and touch the damaged area or distressed person. On its most primitive level, therapeutic touch can be considered a reflex action controlled by the unconscious mind. Many animals show this reflex action to injury, and some of our not-too-distant cousins, the apes, show empathy by touching and stroking others in their group. However, humans have developed a unique, and sometimes complex, ability to formulate methods of caring using touch as therapy, such as massage and reflexology.

The obvious benefits gained from touch were noted by ancient physicians who developed the various touch therapies to a point where distinct disciplines were formed. However, the mechanisms underlying the beneficial effects of the various therapies were largely unknown to their ancient originators. Elaborate theories grew out of this inability to explain the effects, many of which are carried forward today and taught as the philosophies or principles behind the touch-based therapies that are still practiced.

PHYSICAL EFFECTS OF TOUCH

The very action of touching another person connects you to that person's energy field. Recent research at the Institute of Neurology, in Britain, has shown that receiving touch from another stimulates the brain far more than if you were to touch yourself.

Treatment, in its broadest sense, can be said to have begun simply by the laying on of hands. It is not unusual to see people unconsciously drop their tense shoulders simply by placing a hand on the shoulder or back. The power of touch is greatly underestimated, especially by many conventional medical practitioners who rely on drugs for most problems, some of which would benefit from a touch-based therapy.

Therapeutic touch is most often associated with problems relating to muscular tension or injury. It is a fact that the muscles benefit greatly from touch, but the effects would be very short-lived if other body systems, such as circulation, were not also affected at the same time.

Touch is an important aspect of bonding among many groups of animals, such as these Japanese macaques.

TOUCH AND BRAIN STIMULATION

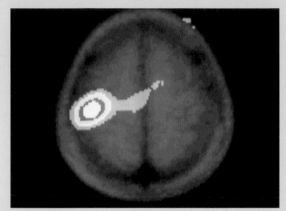

Courtesy of the Institute of Neurology

There is a fundamental difference in our reaction to another person's touch and the effect of self-touch. Such a powerful and positive healing energy is generated between the giver and receiver that the brain shows a profoundly different pattern of brain activity. This will come as no surprise to anyone who has had massage therapy. This PET (Positron Emission Tomography; a form of tomography used for brain scans) scan shows increased activity in the cortex of the left hemisphere during touch stimulation of the right hand.

Effects on the circulatory system

In general, the circulatory system of blood vessels (veins, arteries, and capillaries) can be either toned by a stimulating touch or calmed by a relaxing touch. There is no point in applying a vigorous touch therapy to an agitated, stressed person with high blood pressure, for example, when a calmer more soothing touch could pacify the overwrought system.

Balancing the circulatory system can benefit the body tissues by allowing them to receive the optimum supply of oxygen and nutrients while effectively removing toxins and other byproducts of their metabolism, which is essential for maintaining general good health.

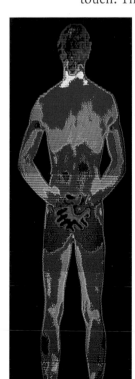

This thermographic image shows the distribution of blood/heat in the body – the coolest areas are dark blue or black, and the hottest areas (that need freeing up) are white.

Effects on the nervous system

Of all the systems affected by touch, the nervous system is the most profoundly influenced. As mentioned earlier, the spinal cord and brain filter process every nerve message that is generated by the application of touch. These tissues react within microseconds. Other nerve tissues, however, may take a little longer to respond to touch. The nerve tissues that form the system in charge of our internal workings, known as the autonomic system, are often out of balance and require a little time to rebalance after the application of therapeutic touch. However, with subsequent applications of touch, the system will respond more quickly. The autonomic nervous system controls those bodily functions that occur involuntarily, such as breathing, heartbeat, blood circulation, digestion, and also the metabolism.

Effects on the lymphatic system

All tissues produce waste products by virtue of living and metabolizing. These toxins must be removed effectively from the body before they start to congest and "pollute." The lymphatic system acts as the drainage network for the body, taking fluids rich in toxins away from the cells and tissues for disposal and elimination from the body, but this system is prone to congestion and disruption. Touch therapy and the movement of tissues assists in the drainage of fluids and the "clearing" of congestion in the system. Once the lymphatic system is "unblocked," it can then function to its full potential and will restore energy to any fatigued and blocked areas of tissue.

Thoracic duct

Lymph vessels

Lymph nodes

The system of lymphatic vessels covers the entire body. Lymph nodes act as infection filters by trapping germs and preventing them from traveling around the body.

Effects on the respiratory system

One of the most visual examples that demonstrates the effects of touch is the deep inhaling breath and often vocal exhaling breath seen when someone receives a touch-based therapy. The body appears to let go of a lot of pent-up negative energy and emotion when a connection is made with someone else's energy field. Soon after the initial contact, breathing can be seen to slow down, and a light, "tidal" flow of inhaling and exhaling breaths replaces the stifled, sometimes poorly coordinated, breathing cycle. Balanced breathing is vital to the regulation of oxygen and carbon dioxide levels in the blood, which, in turn, is needed for the healthy workings of the tissues and brain.

Effects on the digestive system

During the digestive process, the autonomic system takes command, but digestive upsets can occur if this system is not working at its balanced best. Diet has a key role to play, but therapeutic touch has the ability to stimulate the nervous energy in the autonomic system to ensure digestion takes place efficiently. Healthy digestion is the key to gaining the optimum nutrition from the food we eat.

Effects on the endocrine system

Touch stimulates many tissues of the body and can help with the release of many hormones. Adrenaline levels tend to drop during touch therapy, while the levels of other hormones associated with relaxation increase. Many women find that touch-based therapies help alleviate some symptoms of the menopause.

Effects on the skin

All the different types of therapeutic touch, from massage and acupressure to reflexology and Rolfing, have to act through the skin. The healing touch efficiently stimulates the circulation and lymphatic drainage of the skin. Efficient lymph drainage eliminates unwanted toxins and congested tissue fluids, which are rapidly replaced with freshly oxygenated blood and nutrients.

After therapeutic touch, the skin is usually left toned and mildly flushed with a healthy glow.

EMOTIONAL EFFECTS OF TOUCH

There is no question that our emotional outlook is altered when we are touched. It is no coincidence that people tend to open up to their therapist and discuss personal and emotional problems during treatment because the therapy is so holistic in approach. Some people may even release quite powerful emotions through laughter or tears. The overall effect is one of both a mental and a physical therapy.

Do not worry if the therapy has this effect on you because it is quite natural.

You may find many emotions are triggered and released when you undergo a touch therapy. Your therapist will be used to this happening so try not to feel inhibited.

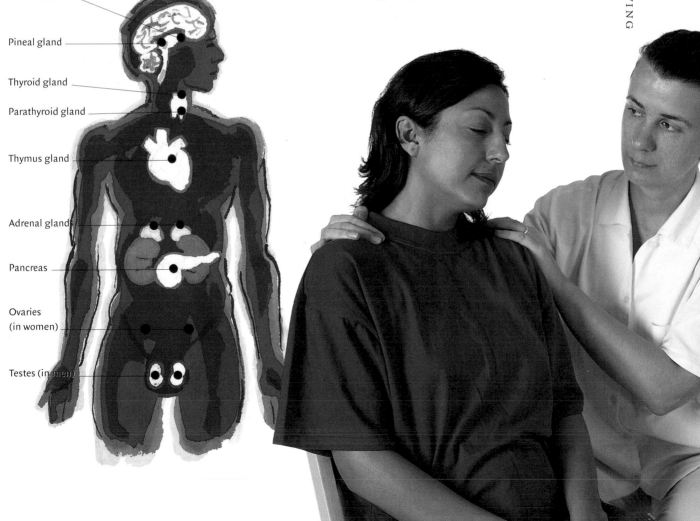

The endocrine glands produce the hormones essential to normal body function.

Pituitary gland

Pineal gland

Thyroid gland

Parathyroid gland

Thymus gland

Adrenal glands

Pancreas

Ovaries (in women)

Testes (in men)

2

TOUCH THERAPIES

The growth of interest in touch therapies is reflected by the number of different types and methods of application available. The choice can be perplexing, especially when each method puts forward convincing arguments as to the advantage of one application of touch over another. This chapter reviews the mainstream therapies available by describing their brief history, principles, and methods of application. It is hoped that this chapter will serve to guide you through this bewildering variety of treatments by allowing you to make an informed decision regarding the type of therapy most applicable to you, your partner's, or your family's needs.

Out of the many therapies described, we have selected four methods that can be safely and effectively used at home without any prior training. The sections describing massage, aromatherapy, acupressure, and reflexology are more descriptive and detailed to facilitate a better understanding of the step-by-step methods used in Chapter 5. However, we must stress that the advice of a healthcare professional should be sought before any treatment whatsoever of an undiagnosed problem is undertaken.

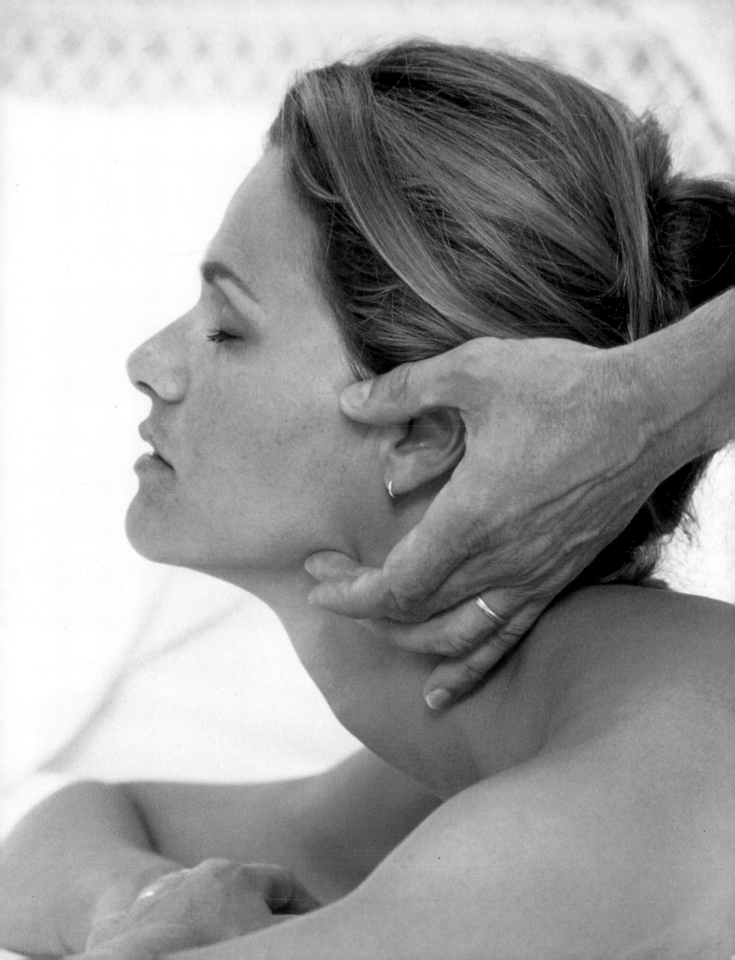

Massage

Humans are sociable animals and enjoy bodily contact. However, in our somewhat clinical and often high-tech environment, the simple laying on of hands can be so absent in our lives that people feel uncomfortable or awkward should it happen. It is interesting to note that in some societies the act of giving massage is considered so important that it forms part of a traditional greeting bestowed on special and honored guests. To this day, the Hawaiian act of lomi centers around this ritual (see page 28).

HISTORY OF MASSAGE

The oldest touch therapy, massage can easily be traced back to around the 5th century B.C. – the days of Hippocrates, the so-called father of modern medicine. In truth, Hippocrates was more of a herbalist and massage therapist. He worked closely with the body using diet, herbs, and physical treatments, such as massage and hydrotherapy, to assist the body to heal itself. Modern conventional medicine has removed itself from some of its original principles.

The history of massage as a promoter of health and well-being is common to most civilized cultures. The ancient Greeks started the first schools of massage where skills were taught in a structured fashion.

It was not until the 16th century that the practice of massage emerged in Europe, but by the 19th century its popularity and demand were rapidly exceeding the number of practitioners who were available to provide it. Every other method of touch therapy has evolved from the ancient practice of massage.

Hippocrates believed that the "cure of disease lies within the body" and that it was essential to understand an individual's symptoms. This remains the cornerstone of holistic medicine today.

VISITING A PRACTITIONER

Before visiting a massage practitioner be aware that you will be asked to undress in preparation for the treatment. However, most practitioners use large towels to cover those parts of the body not being treated and work systematically so no one area is exposed for any great length of time. Towels give an added feeling of security and extra warmth during the treatment and are not just in place for modesty!

A short symptom history is taken before treatment commences if massage is being given for therapeutic reasons. This type of massage is commonly known as remedial massage since it aims to remedy a specific problem. However, massage is generally undertaken to improve your sense of well-being and give you a feeling of increased relaxation.

There can be many variations on a standard massage theme with each practitioner adopting a certain style and medium of his or her own. Oils are commonly used as an effective massage medium closely followed by creams. Some practitioners even use talcum powder. If you have sensitive skin always ask about the massage medium before the treatment begins and, if you are in any doubt about it, dab a small amount on the back of your hand to see if your skin reacts to it and noticeably reddens. If it does, ask your practitioner to use a different medium on you.

Seek professional advice before embarking on any massage therapy if you are affected by any of the following conditions:

- pregnancy
- cancer
- epilepsy
- severe back problems
- severe skin problems
- fever
- HIV and AIDS
- varicose veins and thrombosis
- serious psychiatric illness

(See page 122 for contraindications.)

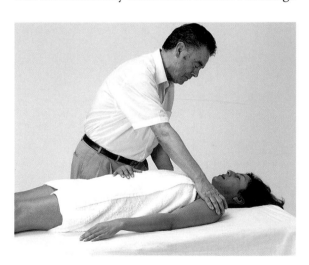

It is important to feel relaxed with your practitioner so that you derive the fullest benefit from the therapy.

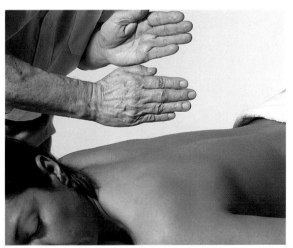

The rhythmic, hacking motion of tapotement is used to increase circulation and tone muscle.

PRINCIPLES OF MASSAGE

Massage is known to affect all body systems, from the emotions, digestion, and circulation through to the nervous system and beyond. However, it is the muscular system that most people think of when massage is discussed. Our muscles form the bulk of our frame, connecting bones to joints and powering every movement from walking to writing. Tension patterns in muscles occur in many predictable locations, such as the upper shoulders and neck regions in desk workers and computer users.

Massage therapy on these areas helps to improve tone and circulation, promoting the elimination of toxins while exerting a balancing effect on the nervous system. Deep relaxation occurs during the treatment, associated with a lowering of the heart and breathing rate, and a calming of the emotions.

The type of massage therapy that will suit you best depends on what you want to achieve. Massage techniques for particular injuries, for example, will be different from those that are more holistic in approach, fostering a general-ized sense of well-being and complete relaxation.

The main types of massage styles are Swedish massage, which uses the four basic strokes shown on the opposite page, Esalen (Swedish with acupressure), lymphatic drainage massage (which concentrates on the elimination of waste products, lomi (a deep tissue massage that is based on a Hawaiian Kahuna tradition), and sports massage (used in training and to treat injury). In Sweden, the massage techniques illustrated on the opposite page are known as Classic Massage.

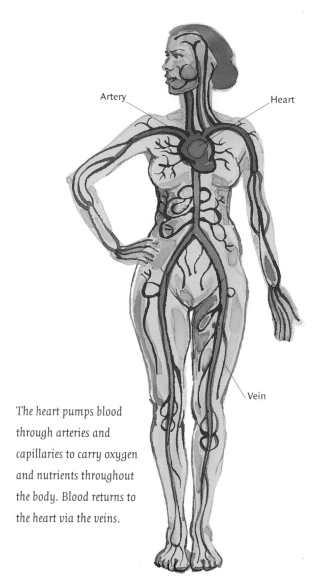

Artery

Heart

Vein

The heart pumps blood through arteries and capillaries to carry oxygen and nutrients throughout the body. Blood returns to the heart via the veins.

A simple 30-minute massage can affect

- the muscular system
- the nervous system
- the lymphatic system
- the cardiovascular system
- the respiratory system
- the digestive system.

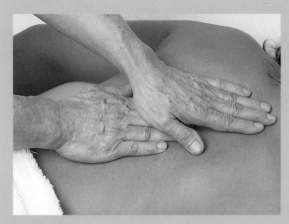

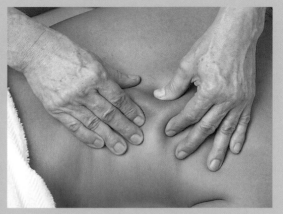

Effleurage

Effleurage is a stroking style of touch that is normally used at the beginning and end of the treatment. It has very soothing properties and must be performed in a rhythmic fashion using both hands in synchronization.

Pétrissage

This is a more vigorous, powerful stroke during which tissues are picked up and rolled. It is suited to very muscular areas of the body, such as the thighs. It has rolling, wringing, and squeezing characteristics.

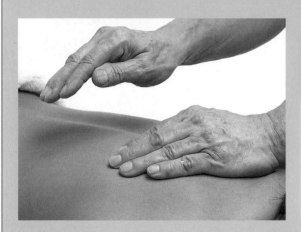

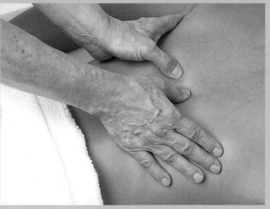

Tapotement

The use of percussionlike movements, using cupped hands or relaxed hands held edge-on to the body, in a rhythmic cupping or hacking action, is probably the best known of all the massage strokes. Tapotement does, however, require practice to perform correctly.

Friction

Using the thumb tips, knuckles, and even elbows, this technique is used to apply a local stimulation to tissues that require special attention, such as knotted shoulder muscles. This technique should be performed firmly, but not so much that it causes any pain.

Aromatherapy massage

Like many of the touch therapies, aromatherapy is not a new art. Combinations of vegetable oils and aromatic plants have been discovered dating back about 4,000 years ago to the Neolithic age. The ancient Egyptians were known to use oil-herb infusions more than 3,000 years ago, and, by 1700 B.C., trade routes that allowed the free passage of aromatic oils and spices had been established throughout the Middle East.

HISTORY OF AROMATHERAPY

Aromatherapy, as we know it today, originated in France in the late 1920s. It was a French chemist, René-Maurice Gattefossé, who accidentally discovered that plunging his burned hand into lavender oil promoted rapid healing and skin regeneration. Being descended from a long line of perfumers, he decided to link the art of perfumery with the science of medicine and pharmacy. The result was aromatherapy, a term Gattefossé used to describe his unique blend of healing art and science.

The sense of smell is an essential aspect of being human. Without knowing it we all use this sense to assist us in making the correct choices. How someone smells has an almost imperceptible influence on how we regard that person. In times past, our sense of smell helped us select ripe fruit and avoid a stomach upset. These days, stores condition us to buy through the use of attractive aromas.

Smell is such an important aspect of life that space scientists have developed bottles of familiar smells to comfort space travelers. This novel approach to homesickness was developed after early moon trips revealed that the lemon-scented moist wipes intended for freshening the face and hands were being regularly sniffed by the crew.

The practise of aromatherapy originated in France early in the 20th century. René-Maurice Gattefossé, a chemist, linked the art of perfumery (shown above) with medicine.

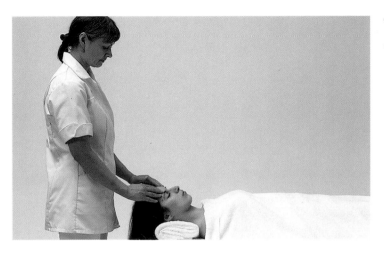

The wrists, along with the temples and neck, are areas that rapidly absorb essential oils.

VISITING AN AROMATHERAPIST

In common with the massage practitioner, aromatherapy masseurs tend to use towels and keep the untreated areas covered up. The added benefit of this practice is to facilitate the effective absorption of the essential oils into the body by keeping the skin temperature warm.

As with other therapies, the practitioner will take a short history relating to general health. He or she will then turn to the important aspect of selecting the correct balance of essential oils for you.

The blend formulated at the end of the consultation is unique to the person being treated. Aspects of your lifestyle and health problems are all taken into consideration during the blending process. A sample of the oil is often given to the patient at the end of a treatment session to use in the bath or to be applied to the wrists, neck, or temple as needed. These areas of the body rapidly absorb essential oils.

After the treatment most aromatherapists recommend that a short rest is taken while toxins are eliminated from the system.

Aromatherapy practitioners

Before anyone can begin to train in the art of aromatherapy, he or she must pass a foundation course in massage theory and technique and, for some courses, must have practiced massage for a period certain of time.

At the heart of good aromatherapy is massage and a sensitive touch. Courses in aromatherapy concentrate on the art and science of essential oils and their safe and effective blending. Like massage practitioners, aromatherapists often work on their own; but many find work in health hydros and spas, and some specialize in the care and well-being of cancer patients, working alongside nurses in hospices.

When you are choosing an aromatherapy practitioner, try if possible to obtain a word-of-mouth recommendation. If you cannot get a personal contact, you may like to go and see the aromatherapist first before committing yourself to a consultation. It is important that you and the practitioner establish a good rapport so that you are able to get the most out of your treatment.

PRINCIPLES OF AROMATHERAPY

There is nothing mysterious about essential oils. Simply squeeze a sprig of fresh rosemary between your fingers and smell the pungent odor that is left behind. The odor you have liberated consists of the essential oils contained within the rosemary leaves. To date, there have been more than 30,000 individual aromatic molecules identified and named with a single oil. This vast and complex array of oils and molecules gives an oil blend its unique characteristics.

Having such a complex chemistry gives apparently unrelated plants similar aromatic properties. For example, plants that have a lemon hint to their scent may originate from very different plant families, but they possess combinations of aromatic molecules that give each a characteristic, but individual, lemon-based smell. This is often referred to as olfactory (smell) shading. Plants with lemon olfactory shading include lemon, lemon verbena, melissa, citronella, eucalyptus, lemon thyme, and palma rosa.

Conversely, there are certain aromatic plants that are made up of only one or two aromatic molecules. These essential oils are unique and demonstrate no olfactory shading. Those essential oils with no olfactory shading include clove and sandalwood.

Using essential oils for health and well-being

The art of aromatherapy hinges around the selection of an ideal blend of essential oils for the problem being addressed. The application of the oil blend is also critical. Oils can be applied directly to the skin in a dilute form, used

The bounty of nature has provided us with a diverse and powerful medicine chest.

in the bath, in a vaporizer, as a steam inhalation, or, under special instruction (not covered within the scope of this book), taken internally.

Skin applications

The highly concentrated nature of essential oils makes the neat application to the skin potentially hazardous, so all essential oils must be diluted in a carrier oil, such as almond oil, before they can be safely used in aromatherapy massage. As a general rule, add 1–2 drops of your chosen essential oil to one tablespoon of

who can benefit from aromatherapy?

People of all ages and lifestyles can benefit from this gentle, luxurious therapy. Children benefit just as much as adults, particularly so just before bedtime because essential oils are absorbed very rapidly. Keep diaper rash away with one of the antibacterial oils. Colic and teething respond well to the relaxant, sedative oils.

carrier oil. Suitable carrier oils include almond, soybean, grapeseed, avocado, and wheat germ.

Once mixed, always store your blend in a dark bottle with an airtight lid. This precaution must be taken to prevent light rays from reducing the potency of the therapeutic actions of the oils by damaging the delicate aromatic oil molecules.

You can use the oil blend as you would use any massage oil and work it well into the skin using your favorite massage stroke. For added benefit, carry a small phial of the oil around with you and dab it behind your ears or on your forehead as needed. These areas of the body absorb the oils rapidly.

Heating essential oils

It can be helpful to take advantage of the ease with which essential oils will vaporize when you use your oil blend as an inhaled treatment. The age-old therapy of steam inhalation can be made more effective by adding a few drops of eucalyptus, the effective decongestant and antiseptic oil, to a steaming bowl.

Bathing with essential oils

Adding 10–15 drops of your favorite essential oil to hot, running water can turn an ordinary bath into a wonderfully uplifting and highly therapeutic experience.

After a long, stressful day, try adding camomile or neroli oil to your bath for a soothing and relaxing soak. Children respond well to a lavender bath just before bedtime and this will help to calm them after the excitements of the day and prepare for bed.

COMMONLY USED AROMATHERAPY OILS

Name of oil	Qualities
Lavender	Antiseptic, antidepressant, and calming
Camomile	Relaxant, anti-inflammatory
Rosemary	Stimulating, pain relieving, decongestant
Lemon	Astringent, antiseptic, antidepressant
Peppermint	Antinausea, stimulating, decongestant, relaxant
Tea tree	Antibacterial, antiseptic
Geranium	Calming, balancing
Lemon verbena	Sedative, digestive
Melissa	Antibacterial, sedative
Citronella	Insecticide, stimulant
Eucalyptus	Antiseptic, anti-inflammatory, decongestant
Lemon thyme	Antibacterial, antiseptic
Palma rose	Antiseptic
Clove	Antinausea, analgesic, antiseptic
Sandalwood	Antidepressant, antibacterial, sedative
Neroli	Antidepressant, antibacterial, antiseptic
Orange	Antidepressant, antibacterial, sedative
Petitgrain/ Pettigrain	Antiseptic, antidepressant.
Clary sage	Antidepressant, stimulant
Mandarin	Relaxant

Acupressure

There can be very few of us who, at some time or another, have not found a tender point in a muscle, given it a press and gained relief. Just as we intuitively rub a painful area, we often find that sustained finger pressure over aching muscles helps to reduce the discomfort and improve freedom of movement. Unwittingly, you have been using a basic form of acupressure. Within the Chinese system of medicine, a common theme of energy balance is at the heart of good health and healing. The vital life energy, known as "chi," is said to flow throughout the whole body.

HISTORY OF ACUPRESSURE

It is difficult to trace the exact history of acupressure since its methods can be seen to be incorporated into other healing arts such as massage, reflexology, and shiatsu. However, the theory and practice have their roots in oriental medicine, and the development of acupressure closely mirrors that of acupuncture, which has been established for many thousands of years. The two treatments are very similar, both in the selection of energy points used and in the diversity of conditions that can be alleviated.

Acupressure describes a large variety of massage techniques that use manual pressure in order to stimulate the energy points on the body, which are associated with the meridians. The pressure exerted is usually light to medium.

Today, acupressure comprises a number of different systems around the world, including shiatsu. Its practice may therefore vary considerably from therapist to therapist.

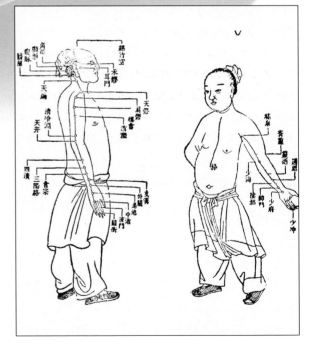

There are approximately 365 acupoints along the 14 meridians of the body. Some are extremely powerful points, while others are weaker. Some points affect a particular body part, while others are more general in their effect.

VISITING A PRACTITIONER

Many complementary health practitioners use acupressure points in their work. You will find that massage, acupuncture, and reflexology practitioners use acupressure most frequently, but other professionals may specialize in this type of energy-point therapy as well.

Acupressure is a very safe form of therapy. It can be used effectively at almost any time of day, by any age group, and by pregnant women.

Treatment is normally given in a similar fashion to traditional massage. You will need to undress to expose the area that requires treatment so that skin-to-skin contact can be made. This is important in energy-point location and energy transfer. Treatments may take from 30 to 60 minutes, depending on the problem being addressed.

It is recommended that, to begin with, weak or fatigued individuals have only short treatments while their vitality adjusts. Pregnant women should also avoid strong stimulation.

Acupressure practitioners

Most acupressure practitioners undertake a foundation training in massage from which they then progress to study specialized courses in acupressure. Some acupressure practitioners started their working lives as acupuncturists who later choose to adopt a noninvasive approach to treatment by replacing needles with finger pressure.

In general, acupressure is a logical development in the professional portfolio of many manual touch therapists. Acupressure also has strong links with other therapies such as reflexology and shiatsu. Like shiatsu, acupressure relies on the thumbs, palms, heels of the hands, and elbows to apply held pressure on vital points on the body.

Acupressure practitioners learn how to locate the pressure points, known as acupoints, very accurately. For an effective treatment the points must be stimulated correctly for relief to be felt and for body energy to flow freely.

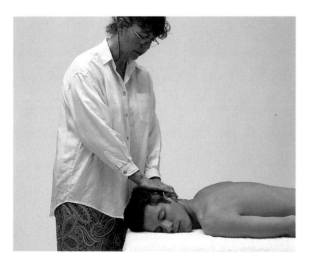

Acupressure uses finger pressure on acupoints throughout the body to stimulate the flow of "chi."

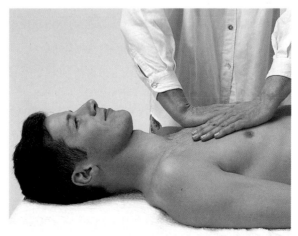

The use of acupressure has been found to be beneficial in the treatment of pain.

PRINCIPLES OF ACUPRESSURE

In good health, the chi is balanced, but in disease or ill-health it may become blocked or congested. This is because it runs along the network of meridians that transport the vital energy to all parts of the body (see page 44). At specific points along these energy lines the energy can be stimulated, either by the insertion of a fine needle (acupuncture) or by the application of finger pressure (acupressure). The practice of acupressure is effective and quite easy to apply, even on yourself.

Acupressure is an extremely versatile method of healing. By selecting the correct combination of energy points, from the 365 points that have been established, treatments can be focused on any body system from digestion through to the nervous system and the brain.

Choosing the correct energy points

There are no set rules governing the selection of energy points. The rule in acupressure is to choose the points that provide the best relief. All the points shown in this book have been well documented as being effective in treating the indicated health problem, but they are not a definitive guide.

Use your own sensitivity and awareness when searching for energy points, and don't be afraid to try combinations that stray from the conventional – they may work for you.

How to apply acupressure

Acupressure should not be a painful treatment in any way, even though the sensations felt can be quite powerful.

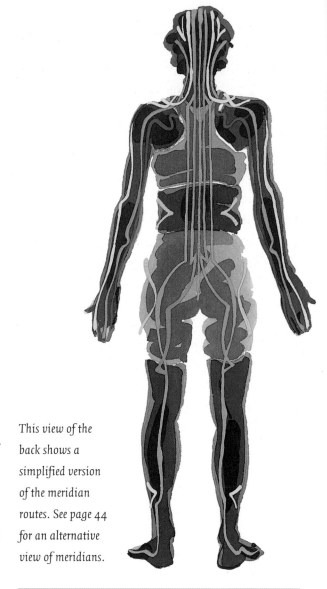

This view of the back shows a simplified version of the meridian routes. See page 44 for an alternative view of meridians.

benefits of acupressure

Becoming familiar with the use of energy-point therapy can be helpful in treating many diverse health problems, as well as maintaining general health and well-being. Acupressure is an effective first-aid treatment for many common ailments, such as cramps, nose bleeds, travel sickness, and toothache.

ACUPRESSURE TECHNIQUES

There are a variety of techniques available to the practitioner once he or she has made a diagnosis. Pressure should be applied gently to start with, before increasing to the point where a strong sensation (not pain) is felt. It is important, however, that the direction of stimulation follows the natural flow of chi in the meridians. Energy points are located using the smooth edge of a finger or thumb. Once located, the energy point can be stimulated using the same finger. This improves energy flow and helps to unblock channels and enhance organ function. With experience, correct levels and duration of pressure become second nature to the practitioner.

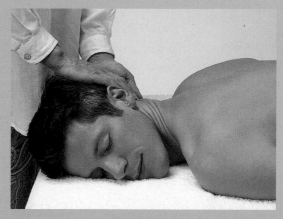

Calming hand technique

When the chi is too active, use the whole palm to gently rub the area concerned. Give a gentle rotating massage to the selected point for between 30 seconds and 2 minutes, keeping a constant level of pressure.

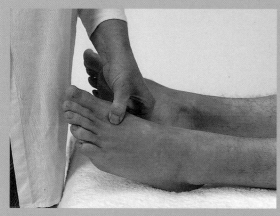

Reducing or dispersing technique

For an excess of blocked chi, apply firm pressure and make small rotations around the selected point (both clockwise and anticlockwise). You can also "pump" pressure in and out of the point.

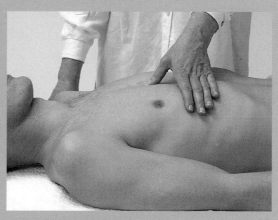

Reinforcing technique

When there is too little chi flowing through the meridians, apply firm pressure to the selected acupoint, without any additional movement with your thumb or middle finger.

Reflexology

It is difficult to dispute the ability of the body to heal itself, given the correct internal environment. Removing blockages that interrupt the healthy flow of energy and stimulating those organs that are deficient or have been starved of energy is paramount to the restoration of health and well-being. In skilled hands, the healing effect of touch can be applied to assist the free flow of energy through the reflex stimulation of channels or meridians. The art of reflexology, or pressure applied to the soles of the feet, is not a new therapy and can be traced back many thousands of years. Reflexology is used to restore and maintain the body's natural equilibrium and to encourage healing.

HISTORY OF REFLEXOLOGY

The earliest evidence of reflexology exists in the form of Egyptian hieroglyphs dating from 3000 B.C. These show Egyptian physicians applying pressure with their hands to the soles of their patients' feet. The hieroglyphs depicting reflexology were drawn alongside other medical procedures, such as childbirth, dentistry, embalming, and circumcision.

The father of modern reflexology, Dr. William Fitzgerald, used pressure points on his patients' feet to induce anesthesia. Another pioneer, Nurse Eunice Ingham, discovered that by using a pressure technique across the soles of the feet, in addition to merely numbing pain she was able to stimulate and heal other parts of the body.

Both concluded that pressure on one area could affect other areas of the body. Therefore, reflex areas on the feet and hands are associated with other areas and organs of the body.

The ancient Egyptians understood the benefits of foot massage as this tomb-painting in the Physician's Tomb in Saqqara testifies. It has also been suggested that the treatment was practiced in China, possibly 5,000 years ago. Practitioners and advocates propose that more than a hundred ailments can be treated by reflexology.

VISITING A PRACTITIONER

On your first visit, your practitioner will give you a preliminary talk. The reflexologist then begins to work on your feet, or hands if necessary, noting problem areas. The reflexologist typically uses touch therapy by applying gentle pressure to the points on the feet associated with the meridians of the body.

A treatment session usually lasts for about one hour and the effect is different for each person. Training sensitizes the hands to detect tiny deposits of waste matter in the feet that can cause imbalances. These deposits often contain lactic acid, uric acid, calcium, and various other organic and inorganic byproducts. By working on these points, reflexologists can release blockages and restore the free flow of energy to the whole body. Tensions are eased, and circulation and toxin elimination is improved.

This gentle therapy encourages the body to heal itself at its own pace, often counteracting a lifetime of misuse. Over a course of treatment, breaking up these deposits by applying therapeutic touch can stimulate circulation and body energy in the area involved.

Occasionally, patients may experience some discomfort, but it is usually fleeting and only indicates congestion or an imbalance in a corresponding part of the body. Generally however, the sensation is pleasant and soothing. Reflexology will relax you while stimulating the body's own healing mechanisms.

About reflexology practitioners

Reflexologists often describe themselves as "catalysts" because they bring about changes in health and well-being, but make no comment on lifestyle or emotional well-being.

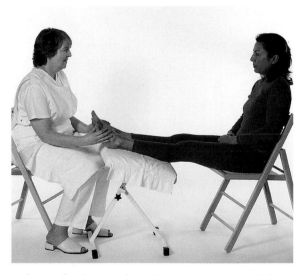

Before a reflexologist makes a diagnosis, he or she will feel for "gritty" or "crunchy" deposits under the skin that are a direct result of imbalances in the body.

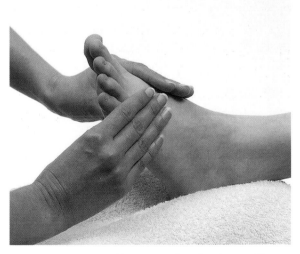

Having located the problem areas, the reflexologist will decide on the course of treatment, working on specific reflex points on the foot.

PRINCIPLES OF REFLEXOLOGY

Reflexology works on the principles of zone therapy to promote health and well-being. In this context the body is viewed as being divided into ten vertical zones running from the feet up to the head and down through the arms into the fingers. It is believed to be through these vertical channels that body energy flows. A reflexology practitioner works on the feet, and sometimes the hands, applying pressure to corresponding organ zones (see adjacent diagram).

By applying massage to the reflex zones, the associated organs are cured of their ailments. For example, locating "crunchy" crystals on the outer side of the big toe suggests that the person is suffering from a bad neck of some kind. By carefully working over this area the problem can be helped and relieved.

In common with acupuncture, reflexology attempts to rebalance the body's healing energy by facilitating the free flow of vital forces and stimulating the healing powers of the body. Reflexologists say that once the energy pathways have been cleared, then true vibrant health can be appreciated.

Although reflexology is itself not harmful, care should be taken by those who suffer from epilepsy or depression. Some practitioners are reluctant to treat anyone with diabetes or women who are pregnant, but there is no evidence to suggest that reflexology can be dangerous to anyone who has any of these conditions. Reflexology is a gentle, noninvasive therapy.

Zones that flow throughout the body can be stimulated by applying pressure to reflex points on the feet.

Who can benefit from reflexology?

Reflexologists say that anyone, regardless of their level of health, could benefit from a session of zone therapy.

From a general health point of view, locating and treating energy blocks before they become serious enough to have an adverse effect on the patient would seem a logical enough reason to consult a reflexologist.

This approach to healthcare closely follows reflexology's own code of practice, which states that its practitioners should only treat the patient's general health and not diagnose medical conditions, prescribe treatment, or claim they can cure any particular ailment.

Many people claim great relief from common conditions such as headache and backache. Stress-related problems have also been treated, and practitioners argue that reflexology can also help in cases of high blood pressure, irritable bowel syndrome, and skin problems such as eczema and psoriasis. Reflexology is unsuitable, however, for anyone who has thrombosis, an unstable pregnancy, or plantar warts.

benefits of reflexology

Reflexology is a holistic therapy and treats the whole person, not the symptoms of disease. For this reason, most people benefit from a course of treatment. The therapy can bring great relief to a wide range of health problems and is safe and suitable for all ages.

THE REFLEXOLOGIST'S MAP

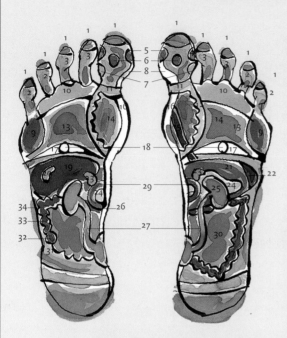

1. Brain
2. Sinuses/Outer ear
3. Sinuses/Inner ear/Eye
4. Temple
5. Pineal/Hypothalamus
6. Pituitary
7. Side of neck
8. Cervical spine (C1–C7)
9. Shoulder/Arm
10. Neck/Helper to Eye, Inner ear, Eustachian tube
11. Neck/Thyroid/Parathyroid/ Tonsils
12. Bronchial/Thyroid helper
13. Chest/Lung
14. Heart
15. Esophagus
16. Thoracic spine (T1–T12)
17. Diaphragm
18. Solar plexus
19. Liver
20. Gallbladder
21. Stomach
22. Spleen
23. Adrenals
24. Pancreas
25. Kidney
26. Waistline
27. Ureter tube
28. Bladder
29. Duodenum
30. Small intestine
31. Appendix

32. Ileocecal valve
33. Ascending colon
34. Hepatic flexure
35. Transverse colon
36. Splenic flexure
37. Descending colon
38. Sigmoid colon
39. Lumbar spine (L1–L5)
40. Sacral spine
41. Coccyx
42. Sciatic nerve
43. Upper jaw/Teeth/Gums
44. Lower jaw/Teeth/Gums
45. Neck/Throat/Tonsils/Thyroid/ Parathyroid
46. Vocal chords
47. Inner ear helper
48. Lymph breast/Chest
49. Chest/Breast/Mammary glands
50. Mid-back
51. Fallopian tubes/Vas deferens/ Seminal vesicle
52. Lymph/Groin
53. Nose
54. Thymus
55. Penis/Vagina
56. Uterus/Prostate
57. Chronic area – Reproductive/ Rectum
58. Leg/Knee/Hip/Lower back helper
59. Hip/Sciatic
60. Ovary/Testes

Every part of the body correlates to a precise area or reflex point on the sole of the foot as well as the areas around the heel and on the foot's surface.

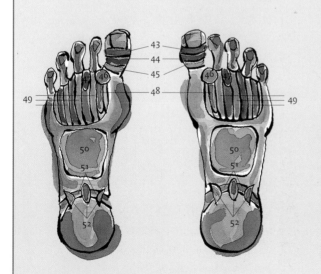

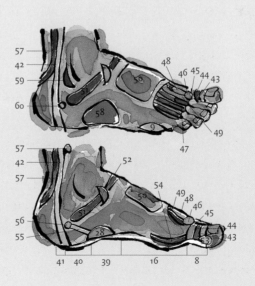

Shiatsu

Shiatsu was originally a Japanese practice of working with the body and, like many other forms of massage, was used to relax and invigorate. The therapy is based on the principles of traditional Oriental therapy and uses acupuncture points to restore the flow of energy through the body to enable a state of balance and well-being. In Japan it is widely practiced by many acupuncturists with an interest in physical therapy. The popularity of shiatsu has grown immensely in the West probably because of its effectiveness in the relief of stress-related problems and its underlying philosophy of stimulating inner well-being.

HISTORY OF SHIATSU

Deriving its meaning from the Chinese for "finger pressure," shiatsu has been used in Oriental medicine since ancient times. Finger-pressure massage was first used in China before being introduced to Japan, but it was in Japan over the last 100 years that greater development of shiatsu occurred, separating it from traditional massage forever. In common with acupuncture and acupressure, shiatsu follows the traditional Oriental concepts of balanced energy flow. However, whereas these treatments do not involve much physical body contact, the great benefits of shiatsu are derived from its highly physical methods. Because the shiatsu practitioner works quickly and energetically, shiatsu has the potential to be more invigorating than other forms of massage and acupressure. Shiatsu is therefore ideally suited to treat chronic fatigue and disease conditions of which fatigue is a major feature.

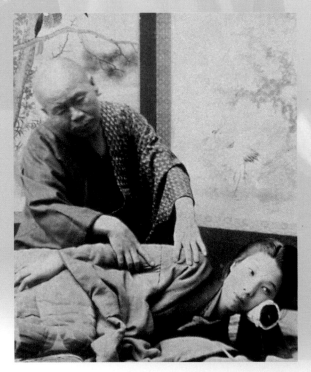

Shiatsu is an accepted medical therapy in Japan today and is used to treat and prevent disease. Some practitioners have incorporated Western elements into their practice so this form of touch therapy can be seen as continually evolving.

Your shiatsu practitioner will take a medical history using methods such as pulse diagnosis.

VISITING A THERAPIST

Shiatsu is truly a whole-body treatment. Sessions normally last up to 90 minutes and require you to lie on a mat placed on the floor. Classic shiatsu is performed fully clothed, with no need to undress. For this reason, it is best to arrive for your treatment in loosely fitting garments, such as a tracksuit.

The practitioner will take a history using Oriental concepts of pulse diagnosis (there are thought to be 28 different pulse qualities indicative of health) and physical examination of the abdomen and muscles. Important areas of muscular tension and energy blockage are determined at this point. During the physical examination and diagnosis, practitioners will interact with their clients. In shiatsu, philosophy, diagnosis, and treatment are said to be one and the same.

Treatment is normally given using pressure applied by the fingers, knuckles, elbows, or knees. Most of the pressure applied during a

session of shiatsu is sustained and concentrates on one area of the body, involving holding and supporting tissues and limbs. This is one of the main differences with the vigorous and mobile traditional massage therapies. Some aspects of shiatsu resemble other touch therapies, such as the similar use of foot-treatment zones in reflexology. The exact style of treatment is dependant on the individual and his or her particular health problem and needs.

Shiatsu practitioners

Shiatsu is normally practiced by those with a good knowledge of Oriental medicine. An in-depth awareness of body energy and energy channels is essential, making Shiatsu treatment a natural development from acupuncture and acupressure. In general, shiatsu practitioners are physically fit and follow a lifestyle in keeping with overall Oriental philosophies for health and well-being.

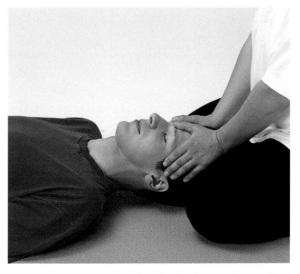

Shiatsu is the application of gentle, but firm pressure on key points of the body.

PRINCIPLES OF SHIATSU

Shiatsu, like many other Oriental therapies, holds that physical and mental illness relates to an imbalance of the body's energy flow, and the therapy aims to release "blocked energy" by increasing, decreasing, or moving the flow of energy that is known as "Ki." This is done by applying pressure from the thumb, finger, elbow, or even the knee along the energy lines or meridians of the body.

The therapy consists of many techniques such as pulling and pushing strokes, tapping, rubbing, stroking, and squeezing so that the pressure applied has an influence upon the tissue beneath the skin. This can be achieved by using pressure and holding techniques combined with gentle stretching.

Regular treatment can help to show any areas of imbalance and by working on them can prevent them from leading on to a more serious condition. For people who already suffer from health problems shiatsu can be of enormous benefit and the therapy will help to alleviate symptoms and prevent the condition from becoming worse.

Many complaints can be treated using shiatsu, but it is especially effective for stress-related symptoms such as tension headaches, muscle tension, and anxiety. However, many other complaints, such as premenstrual tension, back pain, insomnia, digestive problems, and circulatory conditions can also be successfully treated using the therapy.

There are certain conditions where shiatsu should not be used, such as active skin disease, fever, or deep-vein thrombosis.

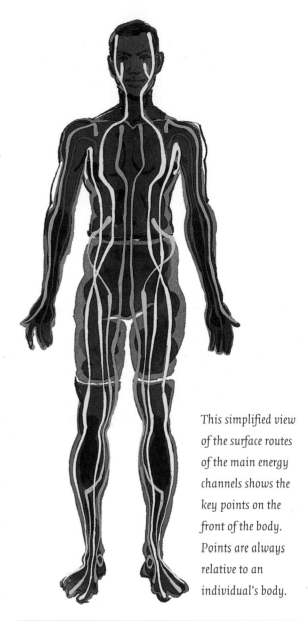

This simplified view of the surface routes of the main energy channels shows the key points on the front of the body. Points are always relative to an individual's body.

benefits of Shiatsu

Shiatsu can help to:
• relieve mental and physical stress
• restore balance and general well-being
• enhance digestive and eliminative function
• ease back pain
• promote good posture.

Palm and thumb

Two of the most basic techniques have been combined here to apply pressure to selected areas of the body.

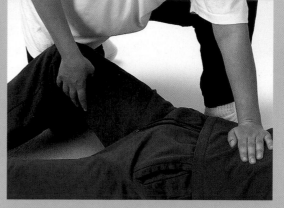

One thumb

In this technique the fingers provide stability as the thumb applies firm and direct pressure.

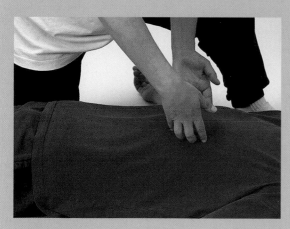

Two hands

Working with two hands, using the knuckles to make contact with the patient, allows the practitioner to closely monitor energy levels.

Elbow

Use of the elbow can offer the practitioner relief from other more strenuous shiatsu techniques, while firm pressure is maintained.

Hellerwork

Hellerwork practitioners say that many common conditions can be helped by the treatment. Not only does this intense therapy make a person more aware of his or her body, it can also help improve posture and relieve aches and pains associated with stress and tension. On a superficial level, Hellerwork would appear to resemble Rolfing. However, its practice and philosophy is designed to affect the body on an emotional, as well as a physical, level. This aspect of Hellerwork differentiates it from the mainly physical and less gentle approach practiced by Rolfers.

HISTORY AND PRINCIPLES OF HELLERWORK

The originator of Hellerwork, Joseph Heller, began his career in touch therapy by training with Ida Rolf in 1972. Heller studied Bioenergetics, Gestalt therapy, Psychosynthesis and Voice Dialogue, and he went on to use this knowledge to develop and expand upon the largely physical concepts of Rolfing. This deep understanding of the emotional and psychological aspects of health allowed the former aerospace engineer to combine them into one complete therapy, which later became known as Hellerwork.

Although Hellerwork is commonly associated with the temporary relief of pain or tension, it was originally developed to rebalance the entire body, returning it to a more aligned, relaxed, and youthful state, helping the patient to become aware of emotional stress that may be related to physical tension.

Former U.S. aerospace engineer Joseph Heller developed Hellerwork in the late 1970s by combining engineering principles and touch therapy to form a new way of looking at and treating the human body.

benefits of Hellerwork

From the philosophical aspect, Hellerwork aims to prevent disease rather than treat it by means of balancing the body and mind so that they work in harmony. However, on a simply physical level, Hellerwork can be very effective in the relief of many common ailments that affect the musculoskeletal system, especially those that have a stress and anxiety source.

VISITING A PRACTITIONER

Hellerwork typically consists of eleven one-hour sessions of deep tissue massage and bodywork, combined with movement education designed to realign the body and release chronic tension and stress. Hellerwork also involves the use of "verbal dialogue," allowing you to become aware of the relationship between your emotions and attitudes and your body. As you become aware of these relationships, you become less likely to limit your body and your self-expression, so freeing your spirit and revitalizing your body. Your practitioner may photograph you before and after treatment to demonstrate how your posture and appearance change over the course of your sessions.

The principal aim of Hellerwork is to realign your body through manipulation to correct problems.

Hellerwork has been likened to peeling an onion – the first layer must be gone through before you can go to the next. To facilitate this layer-by-layer philosophy, Hellerworkers treat the superficial areas first. These comprise the connective tissues that are associated with the muscles.

Following this, the "core sections" are attended to, which are made up of the deep muscles and connective tissue of the body involved with the fine actions needed for graceful and fluid movement.

Finally, the so-called integrative sections are dealt with, whereby the practitioner balances and aligns the unique patterns of each client's body.

By treating the layers of the body – first, the connective tissues, then the deep muscle tissue, and finally the balance of the overall body patterns – successively in this way, very deep-seated areas of tension and rigidity can be effectively treated, thus releasing chronic physical stresses. Several treatments are likely to produce a wonderful sense of renewed energy in the client.

Hellerwork can usually be adapted to suit most people's needs, although it is not suitable for people with some forms of cancer.

About Hellerwork practitioners

Hellerwork practitioners all receive extensive training in the practice and philosophy of Hellerwork and, as individuals, they are committed to the principles of Hellerwork, not only in their actual Hellerwork sessions but also in their personal lives.

Rolfing

According to the teachings of Rolfing, poor posture affects our emotional and physical well-being. This results in a disorganized body that in turn causes more energy to be expended in the maintenance of an upright posture. As time passes, the body tires because energy is used less efficiently compared to someone with a well-balanced body and posture. Rolfing aims to realign the body by manipulating key muscles and connective tissues, letting the body correct itself and recover a sense of balance. It is a very physical therapy and can sometimes be uncomfortable.

HISTORY AND PRINCIPLES OF ROLFING

Since the advent of the Rolf method, Ida Rolf has passed on her knowledge and techniques to many therapists around the world. The number of qualified Rolfers has been estimated at around 700 in total. Her skillful teachings have allowed practitioners of manual medicine to experience the benefits of maintaining a patient's structural integrity because the manipulation of body tissues allows the lengthening of previously contracted and shortened muscles, fasciae, and other connective tissues.

Rolfing's aim is to realign the body by manipulating muscles and other connective tissues in order to enable the body to move more freely and recover its balance. The mind, organs, and other body parts are then able to function efficiently and in harmony with one another, and the immune system can consequently work more effectively.

The system of rolfing developed by Ida Rolf, uses manual medicine to improve a patient's poor posture.

benefits of Rolfing

Rolfing does not claim to cure diseases, but practitioners say that problems affecting the body's skeleton and muscles can be greatly helped by a session. Ailments ranging from backache, constipation, and period pains, have all been helped by Rolfers. Although it can be painful, there is no evidence of harm resulting from Rolfing. However it is not ideal for those who bruise easily or have a low pain threshold.

VISITING A PRACTITIONER

Rolfers are interested in the alignment of the human frame. At every visit the practitioner will assess the balance of the body by examining certain areas of the skeleton, such as the pelvis, leg length, and the tension held in the muscles and fasciae (special connective tissues). Rolfers focus their work on the fascia of the body, which surrounds all the body tissues.

Rolfing is normally given over a course of ten sessions, so it does require a serious commitment on behalf of the client. Rolfers use this time to integrate the Rolfing changes at their own speed, guided by the changes felt in the body's connective tissues at subsequent visits. During a treatment session, the touch of a Rolfer's hands softens, releases, lengthens, and stretches the fascia in appropriate directions. Each session will last about one and a quarter hours, and the course can be completed one per week or spread over as long as six months.

Problems of a physical or emotional nature may arise throughout the course of treatment, causing clients to feel worse before they feel better and improve, but well-trained Rolfers will be able to deal effectively with any such occurrences.

About Rolfers

Rolfers are trained to look at balance in the body as a whole – upper to lower, front to back, inner to outer, side to side, and the overall movement of the human frame.

Throughout the series of ten sessions, they aim to change the body by releasing abnormal or inappropriate tissue tensions in order to restore a healthy balance. This may mean that, for one person, curves in the spine will return, while, for another, the spine may become longer and straighter, having previously contained an exaggerated curve.

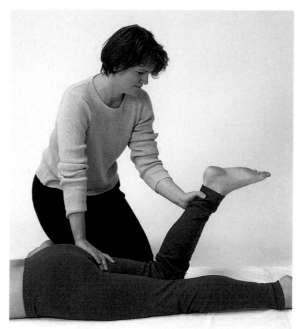

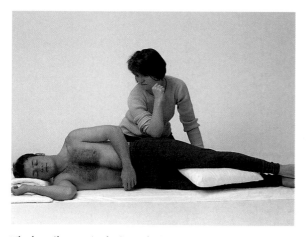

The heavily manipulative techniques used by Rolfers can be quite uncomfortable.

A Rolfer helps to realign and rebalance poor body structure after years of misalignment.

Reiki

Healing and touch have always been closely associated. The art of reiki healing has taken this relationship to new levels. Reiki is a Japanese word meaning "healing" or, more specifically, rei meaning "universal" or "spiritual," and ki meaning "life-force energy." Reiki can help many different ailments, both physical and mental, but its most profound effect is an almost immediate feeling of relaxation, resulting in an enormous reduction of stress, especially if you are feeling low.

HISTORY AND PRINCIPLES OF REIKI

Reiki started in the late 1880s with Dr. Mikao Usul, a Christian minister working in Japan who had been interested in the ancient healing arts. During his studies, Dr. Usul discovered ancient texts written more than 2,500 years ago that described a formula for healing. The text was incomplete but he worked through the writings and revived the reiki process. The same teachings are now practiced worldwide by many thousands of reiki masters and practitioners.

Practitioner–patient interaction lies at the heart of a reiki treatment. Even though many reiki practitioners work with their patients fully clothed, this does not impede an effective treatment outcome. Some even find that powerful treatments can be given without actually touching the patient. A reiki practitioner undergoes special training in order to be able to channel healing energy from the surrounding universe, through his or her own body, into the patient. This effective delivery of the universal healing power separates reiki treatment from all other forms of energy healing. Reiki energy can even be used to heal someone by projecting the energy into the future or to a distant place. Some practitioners have been self-healed by reiki, a process that prompted them to practice the art themselves for the benefit of others.

Reiki promotes health and aims to help clients achieve a state of higher consciousness.

benefits of reiki

Anyone can benefit from this treatment. Reiki is intended to promote all aspects of well-being with the main focus being directed at encouraging the healing energy to flow from the healer's own body to the sufferer. Many conditions that are characterized by deficient energy can gain great benefit. People suffering from low spirits and other emotional upsets often gain the most from reiki. Many physical problems are connected to strong emotions, making them responsive to reiki treatment.

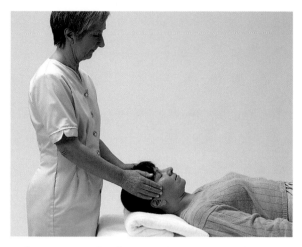

Reiki is an extremely relaxing and calming form of touch therapy, ideal if you are anxious or stressed.

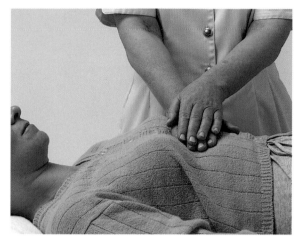

A reiki practitioner will channel a flow of energy to help replenish the patient's energy levels.

VISITING A REIKI PRACTITIONER

A reiki treatment consists of the practitioner placing his or her hands on the client's body, with the intention of promoting energy flow. In a reiki healing session, the practitioner's hands are placed at a number of strategic points. The patient almost always feels the energy begin to flow into the body, either as a heat, a coldness, or a strange "flow" through the body, often in places remote from the point at which the practitioner has positioned his or her hands. Reiki is said to "know" where it is needed and flows to the point where it can do most good. This means there is no complicated ritual to perform – only the simple exchange of energy that occurs between two or more people through touch. In reiki, less is more, and it is often better for the practitioner to do nothing other than allow the energy to do its work.

There is, however, a treatment protocol traditionally taught to reiki practitioners that involves a series of hand positions. Well-spaced along the client's body, these positions provide good and thorough coverage. While the energy does go where it is most needed, it is frequently observed that it stays near where the practitioner has placed his or her hands. Reiki healing is a very safe therapy and is suitable for subjects of any age group.

About reiki practitioners

Reiki practitioners are said to be "attuned" to the reiki energy and continually develop their healing abilities. Their experience and qualifications are graded into three levels.

A Level 1 practitioner is said to be connected to the world of universal energy. A Level 2 practitioner has developed a deeper understanding of positive energy and how to enhance the healing atmosphere that exists at home or at work. When a practitioner reaches Level 3, he or she is attuned to the energy of the universe and is termed a reiki master. At this Level, the practitioner is able to pass on his or her skills to reiki students.

Healing

The human ability to heal others frequently involves the laying on of hands. It is generally accepted that we all have potential power within us to heal others, but some people are able to focus this gift better than others. Healers help others by activating their own internal healing process through the healers' ability to channel energy. No one knows how healing works because it is inextricably mixed with religious belief, which in itself can be considered to be a form of therapy. The healers themselves do not know where their healing energy comes from, but there are plenty of level-headed people who can confirm that the laying on of hands is a pleasurable experience associated with the generation of heat.

THE PRINCIPLES OF HEALING

Healing became an organized therapy in Britain when the National Federation of Spiritual Healers was formed by Harry Edwards in 1955. Edwards regarded himself as a medium for the healing work of Louis Pasteur and Joseph Lister, who, he claimed, worked "through" him. The British Medical Association was not impressed, but it did conclude that recoveries took place through the laying on of hands that could not be explained by medical science.

Therapeutic touch can be considered a modern version of healing. In the United States, Dolores Krieger, a professor of nursing, started the first movement toward therapeutic touch, characterized by the laying on of hands in the fashion of traditional healers, and studied the therapy under scientifically controlled conditions. Therapeutic touch has spread widely throughout the world.

Thought for centuries to be miraculous or "mythical," healing is once again gaining respectability.

benefits of healing

Healing appears to be effective for a wide range of health problems. The power of healers seems to conquer pain of both an acute and chronic nature, as well as reversing disease processes that have been unsuccessfully treated by conventional methods. People attend healers for anything from warts to cancer with variable results. If after two or three visits you feel or see no obvious benefit, healing may not suit you.

VISITING A HEALER

The most common conception of a healer is that of a spiritually linked faith healer. The act of healing is probably one of the safest forms of therapy available since it involves the simple laying on of hands, positive thought, and meditation. The healer merely acts as a channel through which energy passes into the body of the recipient. The transferred energy stimulates the internal healing forces that we all possess but may have temporarily lost.

During treatment, clients may experience a tingling sensation, sometimes called "pins and needles," or may feel heat or a draft coming from the healer's hands.

Genuine healers do not claim to be able to cure disease but insist that they help mobilize the body's healing capacity. Currently open to debate is whether healing works through the concentration of energy, spiritual healing, or by some other means such as positive suggestion or stimulation of the immune system, which would lead to an improved outlook after a healing session.

In any case, healing is an extremely safe and effective therapy for a wide range of apparently incurable conditions.

About healers

There is no special training involved in healing due to the simple fact that those with the gift of healing are often unaware of it until they attempt to use it.

Schools of healing exist not to train healers but to encourage those with the ability to use their energy effectively and efficiently. By virtue of the mental concentration involved, many healers limit the number of healing sessions they perform per day in order to conserve their own energy and remain healthy themselves. Therapeutic touch, on the other hand, is taught to the nursing profession.

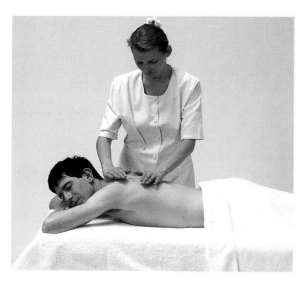

More and more people are turning to healing as a form of preventive medicine.

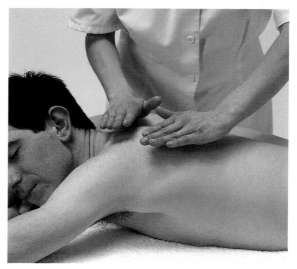

Most healers work with their hands, concentrating on the energy fields surrounding the client's body.

Chinese massage

Massage therapy forms one element of the wealth of knowledge accumulated by practitioners of Traditional Chinese Medicine (TCM). Within China itself the established medical professions are starting to turn the clock back and to look carefully at their medicinal roots. Treatments using TCM methods, including acupuncture, herbalism, and massage therapy, are now finding their way into mainstream healthcare as many people become aware of the side effects associated with modern drug therapies. Chinese massage therapy is becoming readily available in the West as interest in proven Traditional Chinese Medicine rapidly grows.

HISTORY OF CHINESE MASSAGE

Traditional Chinese medicine (TCM) dates back over three millennia to 200 B.C. An ancient text, the "Yellow Emperor's Classic of Internal Medicine," recorded the belief that everything in the universe has a mixture of yin and yang, which must be in balance with each other. This belief is still adhered to by TCM practitioners. The experience and teachings of Chinese physicians has been well documented, with an accumulation of medical knowledge unsurpassed by any other civilization. Massage therapy holds a special place in TCM with treatments that include manipulation, pressing, rubbing, and kneading various areas of the body for the alleviation of internal conditions. Since massage therapy works very well as a general method for relaxation and promotion of well-being, it can be viewed as suitable for specific problems as well as profoundly balancing for general health.

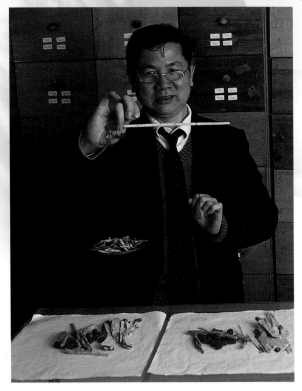

Traditional Chinese medicine, practiced as far back as 2000 B.C., is still popular worldwide.

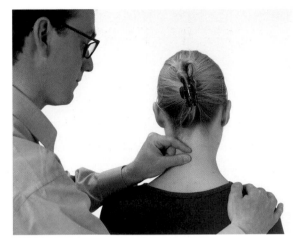

Each massage therapy is tailored by the practitioner to suit your individual needs.

The Chinese massage practitioner will check before treatment begins for areas of built up tension.

VISITING A TCM PRACTITIONER

Care should be taken when selecting a traditional Chinese massage practitioner. It is essential that he or she will have studied many aspects of TCM in order to gain a deep insight into the philosophical aspects of Chinese healthcare and methods.

During your first visit, the practitioner will ask detailed questions about your health and well-being as part of the holistic treatment. She may ask for a full medical history, together with details of your family's medical history. Once the initial consultation is complete, he or she may also offer you guidelines regarding your diet and lifestyle and may suggest herbal remedies, to be used at home between consultations, for any particular problems that you may have.

The elimination of toxins is a cornerstone of TCM and, indeed, of Chinese massage. Your practitioner may therefore wish to work on the digestive and eliminative organs of the body. They may also suggest a detox diet.

At the heart of Chinese massage therapy is individuality: every patient has a unique combination of characteristics, and thus the force and duration of treatment varies considerably from person to person.

Different regions of China follow different guidelines for massage treatment, with some advocating that a particular body site should be massaged up to 300 times in a 25-minute session. Usually a complete Chinese massage takes only a half hour.

About TCM practitioners

Practitioners of TCM may have originally trained in Western methods. Specialist training courses run by Chinese physicians visiting the West on lecture tours offer students the chance to study traditional methods and learn about the TCM philosophy. Most other Chinese massage therapists would have studied on a foundation course in TCM before specializing in the physical modality of massage therapy.

PRINCIPLES OF CHINESE MASSAGE

According to the philosophy of Traditional Chinese Medicine, the body's vital energies, yin and yang, must be in balance for health to prevail. Disease occurs when yin predominates over yang, or vice versa. The resultant imbalance in body energy profoundly affects the functioning of all the major organs of the body, causing malfunctioning and illness. Treatment is aimed at restoring this delicate balance.

The use of Chinese massage techniques has a reflex action on the body's nervous system. In terms of yin and yang, the nerve energy is balanced by causing either excitation or inhibition of nerve activity, which in turn produces the desired medical effect. A good example is the traditional Chinese massage treatment for headache: by applying massage or pressure over the acupoint that is known as Hegu (Large Intestine 24) pain can be rapidly dispelled.

This form of treatment is known as the "pain shifting" method. Chinese massage therapy helps to reduce high blood pressure by causing an opening up of blocked or dilated blood vessels. This effect is thought to be mediated through nerve reflex action. Massage that is applied in this way is believed to have a regulatory action and is attributed to suppressing the liver's level of yang.

Chinese massage therapy was developed to function as an entire system of healthcare. In TCM all illnesses can benefit from the holistic approach at the core of Chinese massage. Chinese massage therapy is particularly suitable for chronic conditions, especially those that affect the musculoskeletal system. However, for the management of acute illness, such as the common cold, joint strain, and slipped disks, many Chinese massage practitioners have claimed to bring about excellent therapeutic results.

COMMONLY USED TRADITIONAL MEDICAL TERMS

Yin: The negative energy associated with contraction

Yang: The positive energy associated with expansion

Acupoint: One of a large number of specific points on the body at which massage, acupressure, or acupuncture is applied

benefits of Chinese massage

Chinese massage therapy helps:
• regulate nerve function
• strengthen the body's resistance to disease
• flush out the tissues
• increase circulation
• make the joints more flexible.

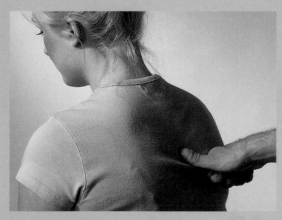

The flat thumb push method

The flat thumb push method is also known as the spiral push method and is used on the back and limbs.

The roll method

This method is best applied to the back, hips, legs, and shoulders, and is often used at the start of the treatment.

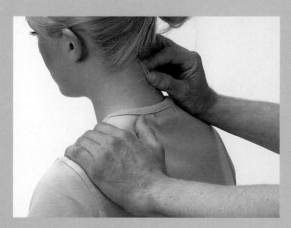

The single finger method

The tip of the finger digs deep into a selected acupoint. This method is most often used on those points located around the head and neck region.

Children's Chinese massage

Chinese massage used specifically to treat children is known as Tui Na. This differs from Chinese massage for adults as development is still in process; for instance, a child's digestive system, unlike an adult's, is constantly working at maximum capacity to ensure the child's growth. Similarly, a child's meridian channels are not fully formed and are slightly different to those of an adult — they have their own set of points which should be established during diagnosis. Treatment for children would include more gentle techniques. For example, by pushing an acupoint or channel, nip-kneading the finger-joints, and stroking the stomach, a child's digestion can be regulated.

Kinesiology

The origins of kinesiology (the study of movement) can be traced back to 1964, when an American chiropractor, George G. Goodheart Jr., first observed that, in the absence of illness or anomaly, postural problems were often associated with muscles that failed special tests designed to assess their strength. Physical treatment of these apparently dysfunctional muscles improved both postural balance and the outcome of later muscle-strength tests. Since then, Goodheart and others have observed that many conservative touch-based treatment methods improved nerve and muscle function to a point where manual muscle testing demonstrated improved strength and ability.

HISTORY OF KINESIOLOGY

The name kinesiology is derived from the Greek word for motion, and the practise uses manual muscle testing to detect imbalances in the body. Along with performing as a diagnostic tool, it has also been developed into a form of therapy by Dr. Goodheart; he uses light pressure massage to acupressure points.

PRINCIPLES OF KINESIOLOGY

Kinesiology, commonly known as applied kinesiology (or "AK"), is a special approach to healthcare that draws together the core elements of the complementary therapies, creating a more unified approach to the diagnosis and treatment of illness.

Kinesiology is often used by manual therapists to test the functional well-being of the body. Assessments are normally used in conjunction with standard methods of diagnosis, such as clinical history, physical examination, and laboratory tests. When appropriate, this clinical impression is used as a guide to the application of various touch therapies.

Kinesiology is also often used as a form of allergy testing whereby a food substance is placed in the mouth and the practitioner tests the resistance of an extended arm. If the arm

who can benefit from kinesiology?

Anyone can benefit from an applied kinesiology examination, and it can be of great value in the management of some common functional health problems such as arthritis. Applied kinesiology helps the practitioner to understand functional aspects of a person's health complaint. When properly performed, applied kinesiology can provide valuable insights into the general and specific well-being of a person.

does not resist, then it is assumed that the person is sensitive to that food. The same principle is used when dealing with psychological problems where the client is asked to think of a certain person or topic while the arm muscles are tested again for resistance which indicates that the person is anxious about that subject.

VISITING A PRACTITIONER

Practitioners of kinesiology aim to assess the health of the whole person, and by using muscle testing they claim to be able to determine the structural and nutritional status of that person.

The sessions can be fairly long, with each visit often lasting an hour as each aspect of your medical history, diet, home and working life are taken into account. The testing and sessions are not painful or dangerous. However, other forms of diagnosis should be used in conjunction with

A kinesiologist assesses the resistance of an extended arm as part of the diagnostic techniques.

the technique. Many patients have claimed that kinesiology has helped them to pinpoint food sensitivities and that they are able to adjust their diet accordingly.

Once an imbalance or allergy is detected, the practitioner will then advise a course of treatment, which may include a special diet, fasting, the use of herbs, and exercise, perhaps in combination with one or two other complementary therapies such as osteopathy or chiropractic.

About kinesiology practitioners

Many kinesiologists also practice some other form of complementary medicine, and in the United States many chiropractors use kinesiology to help support their diagnosis. Along with using kinesiology as a diagnostic tool, other therapists use it to restore energy flow and muscle balance and to help dietary problems by identifying food allergies. While kinesiology is unregulated in Britain, it may be possible to find a practitioner on the recommendation of another complementary therapist.

How does AK work?

There are various factors to the AK approach. Among these is specific joint manipulation or mobilization, various myofascial therapies, cranial techniques, meridian therapy, clinical nutrition, dietary management, and various reflex procedures.

Often the indication of dysfunction is the failure of a muscle to perform properly during the muscle testing procedure. This may be due to an improper balance of nerve and muscular functions.

3

TOUCH THERAPY FOR WELL-BEING

Modern living tends to throw many challenges our way. If we are not careful, the stresses of life can take their toll and health can start to suffer. Taking time to look after your body is an important aspect of modern living, one that many of us do not take seriously enough. In this chapter, we look at the fundamentals of touch therapy and describe how to get the most from its application to general health matters.

In order for therapeutic touch to be effective, it needs to be given in the appropriate environment. This can be easily achieved at home by following the simple guidelines and suggestions outlined. As well as preparing the room, it is vital that you prepare yourself. Applying touch therapy without first warming up can result in problems. Your own muscles and joints require priming in order to function correctly and to avoid injury. Disciplining yourself to use pretreatment warmup will allow you to enjoy the application of touch treatment without the worry of injury or overfatigue.

The main emphasis of this chapter, however, is to offer an effective set of touch treatments to stimulate vitality, relaxation, and sensuality. Giving these aspects of life some time and attention will greatly benefit your general well-being and outlook.

Setting the scene

Touch is just one of our five senses and, to benefit from the healing and therapeutic aspects of touch, the remaining four senses must be in harmony. Setting the scene and creating a safe, warm, relaxing environment is all-important. Whether you are giving or receiving any form of touch therapy, you should also prepare yourself so that you feel completely relaxed and comfortable in order to obtain the maximum benefit from the treatment.

When preparing the room for touch therapy at home, close the drapes and dim the lights a little to calm the eyes and suggest relaxation. Using candles can help to create a special feeling – their flickering flames and light aroma further contribute to the relaxed atmosphere.

Where possible, try to minimize noise and potential interruptions such as telephone calls and children. Choose your time carefully to avoid distractions and to make sure you gain full benefit from the treatment. Provided you are not too tired, early evening is ideal because it is the natural time to "wind down."

Our sense of hearing may find absolute quiet rather intimidating. Having soft background sound can distract from this. The body can often be soothed by sounds from nature, such as waves, rain, or waterfalls. Ethnic pipe music or soft harmonizing voices have a similar effect.

To complete the scene, try using an oil burner primed with a relaxing blend of essential oils, such as lavender or neroli. The power of smell is such that once you associate a certain blend of

oils with a relaxed, warm, and safe environment, the body will immediately react when these oils are smelled again, returning it to a state of calm and peace.

The following practical suggestions are simple to carry out in the home setting. They are designed to provide extra atmosphere and should help promote an effective healing environment in which you can experiment with the techniques outlined on pages 70–119.

Light aromatherapy candles or burn a few drops of a relaxing essential oil, such as lavender or sandalwood, in a diffuser to help create a relaxed and calm atmosphere.

PREPARING YOURSELF

Any discussion with a professional therapist will provide confirmation that giving a good and effective treatment is dependent on feeling physically and mentally balanced yourself. Any tensions or emotional upsets you may be harboring can easily be felt by your subject, who will respond to your "bad vibes." Following a simple ritual of self-preparation just prior to giving a treatment will help to ensure that you minimize any negative energy and maximize the potential healing power of your treatment.

To offer effective touch therapy, you must not feel restricted by clothing, so select free-moving natural fabrics in muted colors and wear comfortable shoes, or remove them altogether. Remove any jewelry that may scratch or interfere with the treatment. Tie or pin back long hair to keep it out of your eyes and away from your subject. Ensure your nails are short and blunt to avoid scratching the person you are treating and wash your hands (see below).

Finally, you need to focus your energy before you are ready to give a touch therapy.

It is sensible to prepare your hands by soaking them in a cleansing mixture of warm water and lemon juice. This helps to soften the skin, stimulate your circulation, and give you added energy prior to contact.

preparing your body and mind

Follow this simple breathing exercise before performing any touch therapy. It will help to focus your energy and boost your circulation, ensuring that your efforts will be concentrated.

1. While the recipient prepares for the therapy, take time to center yourself and focus on your breathing. While standing, inhale deeply through your nose, allowing your chest to expand.

2. Slowly raise your hands over your head and allow the breath to travel slowly down to the abdomen.

3. Hold your breath and sustain the stretch for a few seconds before slowly breathing out through the mouth and lowering your arms. Repeat this sequence three times. You should now be adequately prepared to give treatment.

Getting to know your body

The stresses and strains of everyday life can easily be banished through the benefits of the healing touch in the comfort of your own home. Choose from aromatherapy, massage, reflexology, and acupressure. First, decide what you, and your partner, need and then you can take it from there.

As you and your partner get to know each other's body, you will recognize the tight knots of tension that need relaxing, the sluggish areas that demand touch for vitality, and you will know, too, when the time is right for the sensual healing touch.

Some of us make the mistake, when we are tense, tired, and sluggish, that we need to be invigorated, to receive the healing touch for vitality. However, it is often the case that what we need is deep relaxation. People who used to thrash up and down the swimming pool, for example, or go and play squash to unwind have discovered that what the body and mind really need is for the stresses and tension of life to be released and for relaxation to be actively encouraged.

The discharge of nervous energy is important, but so too is relaxation so the body's cells can rest and repair themselves. This healing process cannot take place if the body is subjected to physical stresses.

For people who habitually feel tired, healing touch for vitality may prove to be of more benefit than relaxing touch. Massage that focuses on lymphatic drainage, for example, may produce the much-needed energy boost.

It is crucial that a practitioner understands the needs of his or her subject; relaxation is as valuable as invigoration.

TREATMENT POSITIONS

Selecting the correct working height is very important. For most cases, working on the floor is adequate. You may find that using a padded mat will make the treatment more comfortable, otherwise use thick towels folded a few times. The careful positioning of cushions can make all the difference. Use a single thickness pillow under the abdomen and feet if your recipient is lying on his or her front, or under the head and behind the knees if your recipient is lying on his or her back. Don't forget to use a small pad under your own knees to avoid discomfort when kneeling beside your recipient.

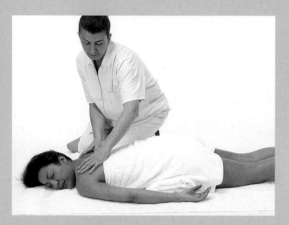

Position for the back

Make sure you lay out towels for comfort and use supportive cushions under the abdomen if you are working on the back. Take time to find a comfortable working posture. Try to kneel to one side of the recipient, keeping your back straight. Swap sides halfway through in order to spread the load on your own spine.

Positions for the neck and shoulders

This area of the body can be easily reached in the sitting position. Try to position your recipient comfortably on the floor or a low stool, sitting cross-legged, if possible. Sit comfortably on a chair behind your recipient and make sure you are well positioned to apply the therapy.

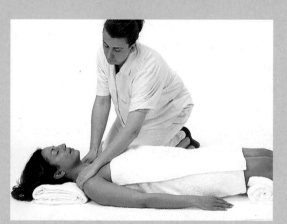

Alternatively, ask your recipient to lie on his or her back on the floor (use supportive towels or cushions as before) while you kneel beside the recipient to carry out the treatment. Again, swap sides halfway through to ensure that you spread the load on your spine.

Touch for relaxation

The increasing pace of modern life has necessitated a need to relax and escape hustle and bustle. Relaxation after a stressful day has never been more important to the maintenance of good health and the prevention of illness. Touch methods, such as massage and acupressure, are very beneficial, helping to release areas of tension and promote a feeling of relaxation and well-being. Massage has many effects, depending on how it is applied. In general, for relaxation, the more flowing and even the movements are, the more relaxing and soothing the massage will feel. Some techniques are better suited than others to this. It is also important to keep your hands and arms loose because this will transmit a feeling of relaxation to your partner. The power should come from your shoulders.

THERAPIES FOR RELAXATION

When most forms of complementary therapies are applied, relaxation often results as a very welcome side effect of the treatment, even if it has been applied for a specific reason such as pain relief. This could be a result of the release from pain or other symptom, but it could also be due to the principle of treating the whole body, including the mind, as is the case in many therapies. Pain naturally exerts stresses upon the body, hindering its natural functions and sapping the body of energy. Pain and stress lead us to make errors, which in turn lead to lack of self-confidence and failure to achieve our goals.

Once the pain is alleviated, the body can relax, energy levels start to rise, and you can once again begin to start functioning to your full potential.

Massage and aromatherapy massage

One of the best-known therapies for inducing relaxation is massage; there are few people who would not benefit from this. When applied using essential oils, such as neroli and mandarin, the effect is even more powerful.

Acupressure

This is particularly beneficial either for yourself or your partner, after a stressful day, for premenstrual tension, or the night before an important meeting or interview.

Reflexology

Once the body is completely relaxed, it is easier to identify those areas that need further treatment. The tender spot located during reflexology massage provides you with the information you need about your partner's vital organs.

This relaxing massage uses the technique of effleurage to stroke away tensions in the body to ease tired muscles, to relax you, and to bring a sense of peace.

Ask your partner to lie on his or her stomach, ensuring that he or she is comfortable (use folded towels for support as necessary).

relaxing essential oils

Add a few drops of one of the following to 1.5 fl.oz. (50ml) of almond carrier oil:

- camomile
- lavender
- neroli
- clary sage
- mandarin
- peppermint

1. Slowly apply long, upward strokes using your palms from the ankle to the top of the thigh. Glide both hands over the hip and slide back to the starting point. Repeat a few times before moving on to the other leg.

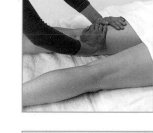

2. Apply the same long, upward strokes from the base of the spine toward the top of the shoulders. Lightly slide down the sides of the body, around the waist, and back to the starting point. Repeat eight times.

3. Slide your thumbs from the base of the scapula – use one thumb then the other, sliding further each time. Repeat three times on each side. Then apply long strokes from mid-spine to the shoulders three times.

4. Gently squeeze the muscle at the top of the shoulders (trapezius) with one hand passing it to the other hand, working the whole muscle. Do each side before stroking again from the mid-spine to the shoulders.

5. Slide your hands lightly down one arm and then stroke from the wrist to the top of the arm. Repeat a few times before sliding your hands across the shoulders to repeat the same strokes to the other arm.

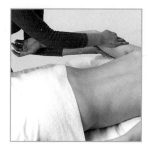

6. Finish by stroking from the top of the spine to the bottom using the fingertips of one hand, followed by the other. Each time allow your strokes to become lighter to ensure deep relaxation.

Touch for vitality

Vitality results from a harmony between the mind and body, allowing energy and enthusiasm to abound. There are few remedies that are as energizing as touch techniques such as massage, reflexology, and acupressure. Touch therapies encourage the recuperation of a tired body as well as an exhausted mind. Many touch therapies are associated with relaxation, in particular massage, but if the techniques are applied at a brisk pace, the effect will be to stimulate and invigorate the body and mind. Certain techniques are more stimulating than others and may be more valuable for those seeking to boost their energy levels and bring an added spring to their step!

THERAPIES FOR VITALITY

Many touch therapies follow the traditional principles of Chinese medicine whose principal aim is to allow energy to flow freely throughout the body. By balancing energy flow through pressure to acupressure points or through the meridians of the body, the recipient often feels energized and full of vitality by the end of the treatment. The Western approach also has an explanation for this release of energy.

Massage and aromatherapy massage

Therapies such as massage free tight knots of muscle that when released give a great feeling of freedom and energy, as well as relief. This is further increased when certain essential oils are added to vitalize the body. Combined with particular massage techniques, such as tapotement (see page 29), this treatment is more likely to increase energy levels.

Acupressure

KIDNEY (K)1: The specific reflex point, Kidney 1, at the center of the sole of the foot, is usually treated to relieve lethargy caused by a build-up of toxins in the body.

K 1

Reflexology

The build-up of toxins and stress areas in the body can be easily located through reflexology. Once these are resolved, energy can again flow freely. When you have completed the treatment, finish with bold upward strokes from heel to toes, ending with a flourish, to banish any last debris.

Shiatsu

Once you are confident of the techniques, apply this therapy for the restoration of vitality. Avoid any tender areas. Shiatsu is a wonderfully invigorating therapy.

stimulating touch therapy

This stimulating massage uses the techniques of tapotement and effleurage across the whole of the body to improve circulation and ease muscle tension, thus giving your subject increased energy and a sense of vitality.

stimulating essential oils

Add a few drops of one of the following to 1.5 fl.oz. (50ml) of almond carrier oil:

- clary sage
- lemongrass
- peppermint
- rosemary

1. With your subject lying face down begin the massage by using long, gliding, upward strokes from the ankle to the upper thigh and over the hip. Increase the pace and repeat the technique three to five times on both sides.

2. Keeping your wrist loose, lightly apply the edge of your right hand so that it bounces off the body. Repeat with the left hand. Apply this alternate chopping movement to the calf and upper thigh. Repeat on both legs.

3. Use the same long, upward movements as in step 1 on the back, starting at the base and moving up to the shoulders. Repeat three to five times.

4. Using your thumbs, apply circular movements to the top of the shoulders, especially where you feel any knots or tension. Repeat the effleurage.

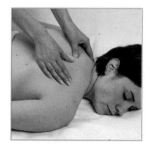

5. Over the base of the spine, where you see dimples either side, apply circular movements with your thumbs in the dimples.

6. With loose wrists, strike your subject's back lightly with the sides of your hands. Use this alternate chopping movement across the whole back from the base of the shoulders, taking care not to apply over the actual spine.

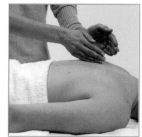

Touch for sensuality

It has been well documented for centuries that massage has played a vital role in intimate relationships. Massage can be the key to successful lovemaking as many couples enjoy the compassionate, unhurried approach that massage offers.

Loss of libido is a common problem and can be due to many causes, including work problems, peer pressure, and lack of self-esteem. Touch is often the most effective method to enhance sensuality and sexual well-being.

THERAPIES FOR SENSUALITY

Sensuality and intimacy are normally based around touch and body contact. For some people, however, this may be an awkward situation to confront and they may need a little help.

First of all, set the scene. You need a warm, well-ventilated bedroom. Have lots of luxurious pillows and cushions around if possible. Subdued warm lighting is ideal, or you may prefer candlelight. Have a pile of soft, freshly laundered towels around so that neither of you need be completely undressed to start with. Start off in soft, silky clothes that are sensual to the touch.

Massage

The use of massage can help overcome any inhibitions and allow the flow of sensual experiences to occur unimpeded. Some may find the use of massage in the sensual environment a challenging experience since it represents a role change from everyday life. Being on the receiving end of massage demands that the subject becomes submissive, extending trust in the giver. This can be a considerable task for many people. However, for the giver of massage, the responsibility for someone else's enjoyment can be equally daunting. The unspoken aspects of massage carry additional sensual messages to your partner, helping bridge the gap between everyday life and the intimacy of the sexual and sensual experience.

Aromatherapy massage

The use of aromatherapy oils in massage can be deliciously sensual, invigorating, and relaxing all at once. Choose the oil most appropriate to your partner from the list on page 33 and apply gently with soft, caressing strokes.

Reflexology

Feet can be very sensitive and sensual, even erotic, making reflexology an ideal choice for the sensual healing touch. Now gradually work your way up the body.

sensual touch therapy

This sensual massage uses slow, gliding movements across the body to relax and arouse your partner. Stroking can be applied to erogenous zones such as the ears, hands, feet, solar plexus, backs of the knees, and back of the neck.

Add a few drops of one of the following to 1.5 fl.oz. (50ml) of almond carrier oil:

- patchouli
- ylang-ylang
- jasmine
- rose
- sandalwood

1 Always begin your massage with slow, rhythmical, even movements to relax your partner.

2 Apply long strokes with both hands, gliding from the shoulders toward the buttocks. Repeat at least five times. This soothing stroking will help to unblock your partner's bladder meridian (important for sexual health).

3 Return to these gliding movements between other movements or changes of position to keep you both relaxed.

4 Massage the lower back. Repeat at least five times before applying long, gliding, stroking movements.

5 Place your thumbs on each side of the spine, apply pressure and hold for a few seconds at regular intervals down the back and across the lower back and hips. Use this technique on your partner's erogenous zones.

6 Now on the front, stroke from under the ribcage down toward the pubis, using one hand and then the other. Using the same technique, stroke clockwise around the umbilicus.

4

TOUCH THERAPY FOR LIFE

Of all the senses, touch is the best developed when we are born. The importance of this can be felt from the moment of birth. The trauma experienced by the baby following entry into the world, full of strange sights and sounds, is immediately calmed when the baby is caressed and cradled by the mother. This mutual touching is as important to the mother as it is to the baby — after nine months they are able to feel each other and, as they become acquainted, the bonding process begins.

Even during pregnancy, mothers instinctively rub or touch their stomachs, even if only briefly, when thinking or talking about their babies. The discomfort many women feel during pregnancy when the baby is lying awkwardly is often relieved by gentle rubbing.

During infancy and childhood, the soothing effects of touch continue more than ever. Just cuddling and rubbing wakeful babies or children is often enough to lull them back into a deep sleep, leaving them feeling reassured and secure that they are loved. When children fall over and hurt themselves, often a brief cuddle and rub to the affected area will ensure that they run off happily and forget their injuries.

Of course, illness can strike at any age, even if it is just a cold or flu, and a great part of recovery is to have a hand held, a foot rubbed, or the brow stroked. Whatever age or life stage we are at, touch as a form of communication should not be underestimated. We use it every day to greet people, to show people that we love them, and to soothe and care for those who have fallen ill. Without touch we feel isolated, alone, and insecure, and even the briefest touch can help lift our spirits.

Pregnancy and childbirth

Throughout the nine months of pregnancy, women can suffer from a wide range of symptoms, but the most common of these are back pain and tension. Both can be effectively alleviated through touch and touch therapies. During early and mid-pregnancy, massage will relax even the most anxious mother. Toward late pregnancy, many women experience back pain as the ligaments supporting the back become lax in preparation for labor. Touch therapy can prove very comforting and can soothe even the most painful back. Always consult your doctor before embarking on any touch therapy (see page 122).

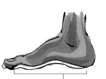

BL23

spine area

THERAPIES FOR PREGNANCY AND LABOR

Many women find pregnancy uncomfortable at some point during their nine months, especially toward the end when back pain can become quite severe.

It is instinctive to hold someone's hand or rub her brow when she is in pain, and for centuries midwives have used touch to comfort and relax women in labor. Increasingly, they are using touch therapies such as massage as a method of pain control.

WARNING
Acupressure Avoid these energy points during pregnancy: Urinary Bladder 60, Large Intestine 4, Spleen 6.
Aromatherapy Potentially toxic oils to be avoided: hyssop, oregano, thuja, bitter almond, pennyroyal, sassafras, and savory.

Reflexology and foot massage
During labor, many women experience cold feet as the body's diverts energy via the circulatory system to help the baby travel down through the birth canal. Yet surprisingly massage of the feet may offer pain relief during childbirth. Specific reflexology points on the sole of the foot relate to the areas most affected.

For example, using your thumbs, gently press all the way up the zone on the foot that relates to the spine. This should relieve tension and ease the pain.

GB 30

Acupressure
As with reflexology, acupressure can offer direct relief. Apply pressure to the following acupoints between contractions: Gallbladder (GB)30, 31; Stomach (ST)36; Bladder (BL)60. Apply pressure to Bladder (BL)60 during contractions.

ST 36

BL 60

touch therapy for pregnancy

Toward middle and late pregnancy, many women find it uncomfortable to lie on their stomachs, so they might find it easier to lie in a fetal position or to sit in a chair leaning over a table, supported by towels or cushions.

Add a few drops of one of the following to 1.5 fl.oz. (50ml) of almond carrier oil:
- camomile
- lavender
- rose
- neroli
- mandarin

1. First mix a few drops of your preferred essential oil with almond carrier oil. Then, using both hands, start at the base of the back and slowly stroke a little of the oil up toward the middle of the back.

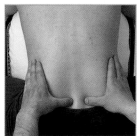

2. Fan your hands and make light, circular, stroking movements, keeping your thumbs on each side of the spine. Repeat this effleurage movement a few times.

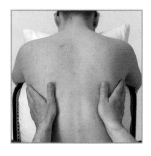

3. Place your thumbs in the "dimples" (sacroiliac joints) in the lower back. Gently press into the dimples, using a slight circular movement.

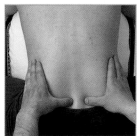

4. When you reach the middle, lightly glide down the sides of the body with both hands and repeat circling up the spine, using the thumbs.

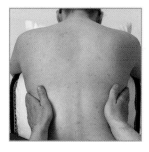

5. Repeat this sequence a few times and then repeat the effleurage.

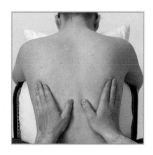

6. To aid relaxation, use thumb pressure on the solar plexus area of each foot.

Babies

It has been well documented that babies who are stroked and touched thrive better than those who are deprived of human contact. From the moment of birth, babies depend on the sense of touch for reassurance and security. Gently massaging babies will enhance the bonding between child and carer, and it can also help calm wakeful babies, allowing deeper levels of sleep. Around 20 per cent of infants suffer from baby colic, causing distress to both parent and child. Massage and aromatherapy can effectively alleviate this painful condition, allowing both to rest.

RELIEVING CONGESTION AND EASING COLIC

Generally, most babies and children, however healthy, will succumb to coughs, colds, and colic. These will lead to congestion and lack of sleep. The symptoms can be eased greatly using massage and aromatherapy massage.

Homemade vaporizer

Add 20 drops of eucalyptus oil to a damp cotton ball, and rest it on a warm radiator. The room will fill with a mild, effective decongestant.

Acupressure

To relieve the sinuses, lay your baby on his or her back. Gently apply pressure with your fingertips around the forehead and cheeks.

Aromatherapy massage

Your baby will probably respond best to aromatherapy massage of all the healing touch therapies. It is gentle, loving, and ensures continued physical contact between you and your baby. Babies naturally seek the reassurance of contact with the parent and may languish without it.

Try aromatherapy massage for any of the following conditions: sleeplessness, crying (when she or he does not need feeding or changing), nervousness, colic, teething, and fever.

Choose from the oils on the table on page 33, remembering to use only one-quarter of the amount that you would use for an adult. Stroke your baby's tummy, arms, and legs and then turn them over to massage her back, and backs of the arms and legs.

> **contraindications**
>
> Avoid massaging your baby if he or she seems unwell or has a fever. Massage should also not be given to babies if they are scarred after an injury.

aromatherapy massage for baby colic

Prepare a warming bath, which will relax your baby before this treatment. For added warmth, you may wish to prepare a hot-water bottle wrapped in a towel.

essential oils for baby colic

Add a few drops of one of the following to 1 tablespoon (15ml) of grapeseed carrier oil:

- dill
- camomile
- lavender

1. Lay your baby on his or her back after a warm bath and spread warmed massage oil very gently over the abdomen.

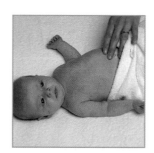

2. With your fingertips, massage around the navel in a clockwise direction.

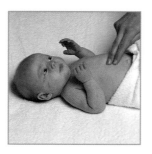

3. Gently press the acupoint Stomach 36 three times. (See pages 36 and 44 for a map of meridians.)

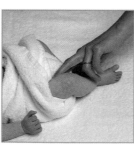

4. With the pad of your index finger, gently press the acupoint Conception Vessel 12 three times.

5. Gently rest your baby stomach-down over a warm (not hot), covered hot-water bottle.

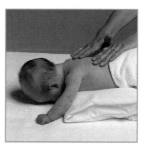

6. Gently and rythmically rub the back. Finish with long sweeping movements again.

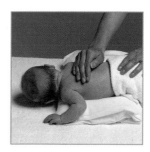

Children

As babies grow into toddlers, suppleness becomes of key importance. Each joint should be encouraged to move on a regular basis so that the child's joints are strong and allow free movement. Inflexibility is rooted in childhood, but it becomes more noticeable in adolescence, when it is often accompanied by bad posture and stiff joints and muscles.

THERAPIES FOR CHILDREN

As children grow older, touch is often offered less. When a child falls, he or she might not receive the soothing rub that would have been offered in earlier years. Sometimes, this is deliberate in order to "toughen them up." However, this can suppress normal feelings and, as the child moves into adolescence and adulthood, can cause difficulties in distinguishing between physical and psychological symptoms.

Adolescence is the most difficult, fraught, and emotional time during life. It is also the time when we have the least physical contact with other people. Adolescents tend to pull away from the reassuring cuddles of parents, although this is also the time when they feel least secure. This lack of touch can lead to a strong feeling of loneliness, low self-esteem, and insecurity.

Massage

Regular massage can aid muscle and joint development. It can also aid bonding and reassurance between older child and parent. Try to incorporate massage (at least the neck and shoulders) into your children's lives as often as possible. Once the second child arrives, the older child may feel neglected and embark on attention-seeking behavior. Massage is a very effective way of showing that you care just as much for the older child as for the new arrival.

Aromatherapy

A simple room vaporizer can ease children who have congestive breathing problems and soothe them into sleep. For a simple remedy try placing 10 drops of lavender oil onto a damp ball of cotton and rest it on a warm radiator. The rising heat will soon fill the room with a calming and sleep-inducing aroma.

You may like to give children an aromatherapy massage if they come to you with problems such as insomnia, toothache, or school phobia! Aromatherapy is also beneficial to allergy, anxiety, bites and stings, bronchitis, burns and scalds, colic, constipation, eczema, and travel sickness. Sometimes, it is lovely to give your child a massage for no particular reason at all.

It is a good idea to introduce massage as a regular feature of your child's life to help keep him or her supple and to help overcome any awkwardness with touch in later years.

relaxing massage oil for children

Mix the following with 3 fl.oz. (100ml) grapeseed carrier oil and keep it in a dark airtight container for when you might need it:

- 6 drops lavender
- 4 drops orange
- 2 drops camomile
- 2 drops ylang-ylang

4

CHILDREN

1. Apply long, upward strokes to the back fanning over the shoulders and down the sides. Repeat a few times.

4. Move on to the arms, gently lifting one and applying long, upward strokes. Give the arm a gentle pull and squeeze the hand before moving to the other arm.

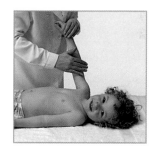

2. With your fingertips, gently circle up the neck to the base of the head, slowly glide down the neck and repeat three times.

5. Use the same long movements for the legs, giving them a slight pull before moving to the feet.

3. Moving on to the shoulders, gently pick up the muscles and pass them from hand to hand (pétrissage).

6. Apply circular movements to the base of the foot with your thumbs before repeating on the other leg and foot.

Adults

It is during the adult years that pressures from life can become overwhelming and have an extremely detrimental impact on our health if pressures persist unchecked. These pressures can include financial worries, excess workload, lack of employment, and problems with family and friends. They can result in a myriad range of symptoms, the most common being tension in the neck and shoulders, headaches, and difficulties in sleeping.

THERAPIES FOR ADULTS

One of the most common problems to beset men and women of all ages is sleeplessness, known as insomnia. Many causes have been identified, but stress and tension remain the most common.

In the bid to look and feel young, many adults undertake some form of exercise. Unfortunately, due to insufficient warming up of the muscles before exercise and cooling down afterward, the likelihood of injury increases. This can be avoided by regular massage.

Aromatherapy massage

A combination of two or three of the following essential oils will be effective for a stress-relieving shoulder or back massage. Add 2 drops of each to some almond carrier oil. Stress-busting oils include clove, basil, camomile, clary sage, jasmine, frankincense, lavender, melissa, neroli, rose, and ylang-ylang.

Just before bedtime try taking a hot bath with 10–15 drops of lavender oil added to the bath. Aromatherapy in the form of a bath infusion can help relax muscles and joints.

Massage using selected essential oils can be of great benefit to those who have insomnia. Try adding 5 drops of lavender, lemon balm, ylang-ylang, or camomile to some almond carrier oil.

Add a few drops of eucalyptus oil to some almond oil to create a massage oil for sports stresses and strains. Alternatively, add a few drops of eucalyptus essential oil to your bath.

Sports massage

Sports activities can result in injuries, especially if the tissues have not been adequately warmed up beforehand. Work with a Remedial Massage practitioner to discover which muscle groups require regular stretching and make sure these are warmed and adequately stretched prior to any sports activity. Self-massage, using a lotion containing eucalyptus and menthol, will help stimulate circulation and muscle heat.

If, however, injuries do occur, then the following actions will help to limit the degree of tissue damage:

- apply ice as soon as possible
- raise the damaged area
- use a compression bandage to limit the spread of swelling, as long as there is no obvious immediate and gross swelling, which will require professional medical attention and, possibly, X-ray investigation
- start gentle massage using arnica cream.

If the problem gets worse or is making slow progress seek professional medical help.

the benefits of sports massage

Massage for exercise will:
- increase flexibility of muscles
- aid recovery of muscles after exercise
- decrease the risk of sprains and strains
- increase circulation and lymph flow
- decrease cramps and stiffness.

ADULTS

relaxing self-massage

This relaxing self-massage will help you unwind after a hard day's work. Take a warm bath or shower to relax help you relax first. Add a few drops of one of the following to 1.5 fl.oz. (50ml) of almond carrier oil: lavender, bergamot, basil, grapefruit, patchouli.

1. Place both hands each side of the neck and slide them from the base of the head down toward the shoulders. Repeat this three times.

2. Then, using the fingertips in a circular movement, slowly work down the neck and across the shoulders.

3. With the right hand, grasp the left shoulder and gently squeeze the muscle. Repeat this three times before changing hands and moving on to the other side. Finish by returning to the long stroking movements from the neck and across the shoulders.

The elderly

With increasing age, many body systems begin to slow down. This is especially true for the circulatory system, and touch therapies, together with exercise, can be beneficial to health. Touch therapy can alleviate other common problems associated with old age, such as stiffness and pain in joints, poor digestion, general fatigue, depression, and anxiety; it can also promote relaxation and increased energy. Most important, touch will decrease the feeling of isolation and insecurity often experienced by the elderly.

THERAPIES FOR THE ELDERLY

One universal aspect of aging is the wearing of the joints. This mechanical process makes them stiff and uncomfortable to live with. Light exercise is essential for general health and is of great benefit to the joints. Regular swimming can develop strength in weak muscles while exercising the joints in a weightless environment. Yoga will also help promote joint mobility as well as improving posture.

When you consider that the heart beats on the average 100,000 times a day, it is no surprise that as it gets older it may need some support in maintaining an effective circulation of blood. Massage is one of the best methods to help boost the circulation. However, take care over inflamed veins and any obvious varicose veins.

Massage and aromatherapy massage

Seeking help from a massage practitioner or aromatherapist should help ease any discomfort felt in the joints. Aromatherapy using the essential oils lavender or camomile is useful, while the more stimulating action of tiger balm may help when applied on a regular basis. You could try using stimulating oils such as black pepper, rosemary, or peppermint to give a boost to a sluggish circulation.

For a nerve tonic mix the following oils and store in an airtight container:
- 8 drops lavender
- 6 drops marjoram
- 4 drops camomile
- 1.5 fl.oz. (50 ml) almond oil.

For a joint tonic mix the following oils and store in a dark, airtight container:
- 8 drops birch
- 6 drops marjoram
- 6 drops lavender
- 4 drops ginger
- 1.5 fl.oz. (50 ml) almond oil.

Reflexology

If you want to improve your general circulation, you should try to massage your hands and feet at least once every day (see right).

relaxing hand and forearm massage

Elderly people tend to suffer from stiff joints; the hands and forearms are particularly prone to stiffness as we use them in so much of what we do. This massage helps to relax joints and muscles and encourage mobility.

Add a few drops of one of the following to 1.5 fl.oz. (50ml) of almond carrier oil:

- 5 drops lavender
- 2 drops frankincense
- 2 drops rosemary
- 20 ml almond oil

1 Spread the oil onto the palms of the hands. Using the thumb of the left hand, knead the right with gentle circular movements.

2 Work around the palm toward the fingers before lightly pulling each finger. Finish off with the stroking action again.

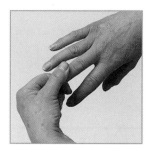

3 Knead the muscles of the forearm to the elbow. Finish with the same long, stroking movements from the hand to the elbow. Repeat the massage with the other hand and arm.

4 Turn your arm and use the same long stroking movements up to the inside of the arm to the elbow.

5 Then, using the thumb and fingers, knead the muscles of the forearm to the elbow. Repeat this a couple of times.

6 Finish with the same long stroking movements from the hand to the elbow. Repeat the massage with the other hand and arm.

5

TOUCH THERAPY FOR BODY AREAS

We have made this part of our guide to therapeutic touch as simple as possible to follow without skipping over important aspects of safety. We recommend that you consult a health professional at all times before treating yourself or anyone else with a health problem or pain that has not been medically examined.

This chapter is divided into body systems, making it easy to locate the area that requires attention. Each body area is approached by the four main touch therapies that can be safely applied at home by yourself or a friend, without any prior training or knowledge. Aspects of aromatherapy, massage, reflexology, and acupressure are included, together with helpful touch tips and lists of useful essential oils and oil blends to help promote the well-being and health of the specific body system in question.

We recommend that you consult Chapter 2 first and read through the descriptions of these therapeutic approaches to gain an insight into their philosophies and application before using the methods outlined here. This will enable you to gain the maximum benefit from the use of touch therapy.

The immune system

An effective immune response is vital to general health and well-being. Those with low vitality tend to fall ill very easily and catch opportunistic infections. In most cases improving the energy levels and boosting the immune system's function can be enough to reestablish an effective defense mechanism and ward off recurrent illness.

The 20th century has witnessed a revolution in medicine and healthcare. Medical research has revealed new facts about the workings of our bodies, both physically and mentally. This has revolutionized the medical profession and benefited all of us in the treatment of ill-health.

One of the most amazing discoveries was that the immune system receives a nerve supply. Until recently, this system of specialized cells concerned with protecting the body from ill-health and invasion by disease-causing bugs was thought to work independently of the brain and nervous system. The discovery that key immune tissues such as the thymus gland and bone marrow (where immune cells grow and develop), as well as lymph glands, receive nerve fibers from the autonomic system (see pages 100–101) suggests a fundamental controlling factor on the immune response – the mind and our emotions. For thousands of years traditional healers have used the phrase "mind, body, and spirit." Medical science now recognizes this relationship and treats the patient accordingly.

Diet, exercise, sleep, and the healing touch are all essential to the healthy functioning of our complex immune system.

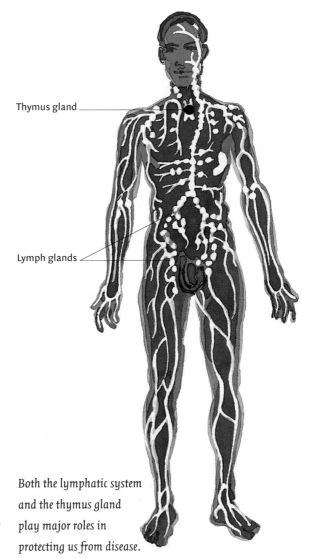

Thymus gland

Lymph glands

Both the lymphatic system and the thymus gland play major roles in protecting us from disease.

STIMULATING THE LYMPHATIC SYSTEM

The lymphatic system carries most of our immune cells from points of infection and acts as a reservoir for our immune reserve. During infection and ill-health, this lifeline can become congested with fluids and the byproducts of disease – toxins and dead cells.

Aromatherapy massage

Stimulating the lymphatic system can have two great benefits: the removal of waste matter (toxins), and the stimulation of lymphatic circulation, in turn aiding the distribution of lymph and immune cells.

Try the aromatherapy massage described here (see box right) to boost the lymphatic system by stimulating the arms. You can also use similar techniques on the legs.

Start at the foot and ankle. Milk the tissues of the calf area up toward the back of the knee: a great deal of accumulated swelling and tissue toxins can be removed in a few firm strokes. Take the back of the knee as your next starting point and push fluids up toward the groin area in order to help drain excessive tissue fluids back into the lymphatic circulation.

BOOSTING THE IMMUNE SYSTEM

In conjunction with a healthy diet and lifestyle, touch therapy can be used to stimulate the immune system in many ways.

It can also help to promote general well-being and help relieve symptoms of fatigue that commonly accompany conditions related to immune-system deficiencies.

lymphatic aromatherapy for arms

Using long sweeping strokes, keep your hand firm and your fingers together, and follow the contours of the arm.

Use the following oil formula:
- 6 drops lemon
- 6 drops grapefruit
- 6 drops pine
- 3 drops thyme
- 2 fl.oz. (60ml) almond oil

1 Working from the subject's hands and forearms, apply a firm pressure between your hands and skin.

2 Milk the tissues upward toward the elbow joint two or three times. Use plenty of oil to help ease the stroke.

3 Take the elbow as your next starting point: milk the upper arms toward the axilla of the armpit. By pushing toward the joint areas, fluids are drained to the lymphatic nodes and vessels where they can be drained away.

Acupressure

For good general health, the immune system needs to function in a balanced fashion. Energy points can be used to boost immunity and should be used on a daily basis as an effective method of keeping strong and healthy when those around you are falling victim to coughs and colds. Apply pressure to the following key acupoints.

GOVERNOR VESSEL (GV)20: Located on top of the head, this point can be stimulated by thumb pressure to boost the immune system and stimulate body energy. Finding this point is quite easy; simply feel the scalp over the apex of the skull for a small tender depression. Once found apply firm pressure over this point for about 30 seconds.

GV 20

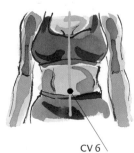

LARGE INTESTINE (LI)4: To increase the immune response, particularly in the upper part of the body, stimulate this powerful acupoint. It is found in the web space between the thumb and first finger. Apply pressure using the thumb of the opposite hand for about 30 seconds.

LI 4

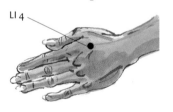

CONCEPTION VESSEL (CV)6: To promote health and energy in the abdominal cavity, especially the detoxifying capacity of the liver and related immune glands,

CV 6

use steady fingertip pressure over this point. The point can be easily found in the middle of the lower abdomen by measuring two fingerwidths down from the bellybutton. Apply pressure here for about 30 seconds.

Reflexology

Try the reflexology exercise opposite to boost the immune system. This is aimed at the chest, thymus, neck, thyroid, tonsils, pituitary, spleen, liver, adrenal, intestinal, kidney, and groin areas of the foot.

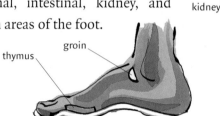

pituitary
neck, thyroid and tonsils
chest
liver
kidney
intestines
spleen
groin
thymus

Aromatherapy massage

Mix together the following essential oils with 1.5 fl. oz. (50ml) of almond carrier oil for a boosting or stimulating combination of aromatic oils to be used during any simple massage of the body (see page 67):

TO BOOST YOUR IMMUNE SYSTEM:

- cinnamon
- lemon
- tea tree
- thyme
- sage
- lavender

TO STIMULATE WHITE BLOOD CELL PRODUCTION:

- bergamot
- myrrh
- sandalwood
- pine
- camomile

Reflexology has been used for many years to boost an ailing immune system. Certain points in the foot can help to enhance and maintain the health of this all-important system and help to prevent the usual infections, which circulate around us on a daily basis, from taking hold. Try using this simple step-by-step treatment during the cold and flu season.

1 Hold both feet and pull both legs down and together, stretching from the hip.

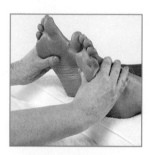

4 Using your thumb on one foot, bend it slightly at the joint so it slides slightly when it straightens. This is known as "thumb walking." Use this technique to walk up the five zones of the foot (see page 41).

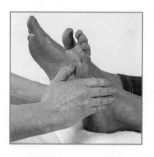

2 Holding one foot at the heel in your hand, slowly and gently move the foot with your other hand so that the ankle moves from side to side. Repeat with the other foot.

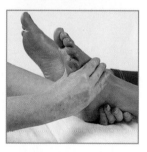

5 Thumb walk over the chest, thymus, neck, thyroid and tonsils, pituitary, spleen, liver, adrenal, intestinal, kidney, and groin areas of the foot.

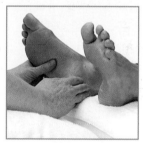

3 Gently press your thumbs on the solar plexus point (see page 41) in each foot and use a slight circular movement on the area.

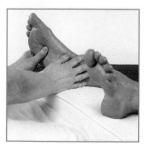

6 End by sliding one thumb after the other, from heel to toe, over the base of the foot.

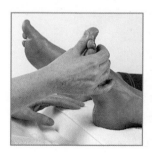

Muscles, bones, and joints

The bones of the skeleton give the body its framework. They are joined together by ligaments, but it is the muscles that lie between the skin and bones that, when contracted, allow the joints to move. There are 206 bones and many more muscles in the body, so it is not surprising that many people suffer from musculoskeletal problems due to bad posture, overuse, underuse, and misuse at some point in their lives. Usually, the result is pain, stiffness, and inflammation of the area, which leads to reduced mobility.

Four main muscle groups are commonly targeted to relieve tension, pain, and stiffness in areas that bear most of the load.

The trapezius is in the upper back and neck and is commonly involved in tension-related problems, causing headaches when very stiff. Problems here arise from general postural defects or poor sitting and working postures.

The latissimus dorsi forms a large sheet of muscle that fills the flanks. It is involved in moving the arms backward and inward toward the body. Problems here can arise from a poor sitting posture as well as inappropriate or unusual exercise involving the arms and trunk.

The deltoid covers the outer part of the shoulder joint with a protective pad of thick muscle. Its main function is to help elevate the shoulders and arms. Pain here is associated with overuse, such as lifting and carrying heavy loads.

The important gluteal (buttock) muscles, the gluteus maximus, medius, and minimus, stabilize the hip joint and lower back. Aches are common in arthritis and sciatic conditions.

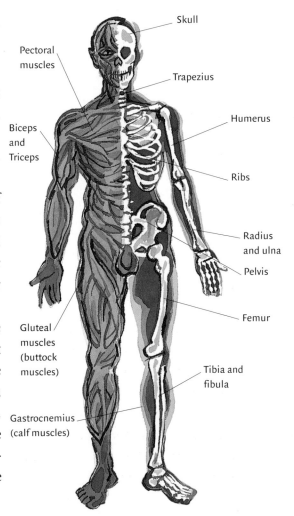

Skull

Pectoral muscles

Trapezius

Humerus

Biceps and Triceps

Ribs

Radius and ulna

Pelvis

Gluteal muscles (buttock muscles)

Femur

Tibia and fibula

Gastrocnemius (calf muscles)

NECK AND SHOULDER ACHES

Neck pain has many causes, but it most usually involves tightening of the muscles surrounding the neck and head. This leads to pain and stiff movement. If left untreated, it can lead to other symptoms, such as headaches, tiredness, back pain, shoulder pain, difficulty sleeping, blurred vision, and feelings of nausea.

Acupressure

Specific energy points can be stimulated to help maintain healthy muscles, bones, and joints. Stimulating the meridians associated with muscles and tendons can help to strengthen and improve flexibility.

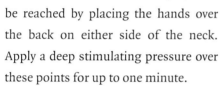

URINARY BLADDER (UB)11: This point can be reached by placing the hands over the back on either side of the neck. Apply a deep stimulating pressure over these points for up to one minute.

SMALL INTESTINE (SI)3: Even though this point is on the outer side of the hand, at the base of the little finger, it can have far-reaching effects on neck pain. Try stimulating this point every hour for 3–4 minutes each time.

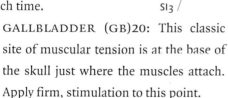

SI3

GALLBLADDER (GB)20: This classic site of muscular tension is at the base of the skull just where the muscles attach. Apply firm, stimulation to this point.

GOVERNOR VESSEL (GV)14: Located between the last bone in the neck and the tip of the shoulder blade, this point should be stimulated deeply by finger pressure to relieve shoulder and neck pains.

GV14

massage for releasing tension

Identify the areas of muscular tension by lightly squeezing the muscles that overlay the back of the neck and shoulders. You will feel areas of tension as knots of "crunchy" tissue that quickly slip away from your fingers when pressed, when your subject may flinch a little! Mix the following oils with 1 tablespoon (15 ml) almond oil:

- rosemary
- camphor
- black pepper

1 Apply some oil to your hands and gently introduce a simple up and down stroke to the neck muscles, working into the upper shoulder and shoulder blades. This starts the relaxation process.

2 With your fingers held closely together, run your thumbs up and along, parallel to the spine while your fingers "pick-up" the upper border of the trapezius muscle. This is a deeply relaxing technique.

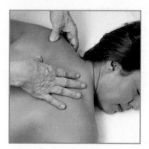

3 End by repeating step 1 interspersed with a fingertip "kneading action" applied to the base of the skull. Releasing tension held in the muscles located at the top of the neck improves head posture and reduces tension headaches.

THERAPIES TO EASE THE BACK

Back pain is one of the most common musculoskeletal problems – around 60 percent of people suffer from it at some point in their lives. If the cause of pain is not resolved, a more chronic condition, such as osteoarthritis (wear and tear of the joints), may develop over time.

Acupressure

Stimulate specific acupoints to help relieve common types of back pain.

URINARY BLADDER (UB)23: For lower back pain. Located at approximately waist level and two fingerwidths away from the spine, this point is commonly tender in back pain sufferers. Apply firm and constant pressure over this point for about 3–4 minutes.

UB23

GALL BLADDER (GB)30: For sciatica. Situated over the area of the sciatic nerve, Gall-bladder 30 is well suited to helping back pain sufferers gain relief from nerve pain related to a back problem. The point is easily found on the sides of the buttocks, two-thirds of the way between the sacrum and tip of the hip. Apply a steady, deep pressure using a supported thumb or elbow with a rhythmic circular action if additional stimulation is needed.

GB30

Massage

Place your thumbs on each side of the spine and apply a slight downward pressure into the back and hold for 4–5 seconds. Move all the way down the back to ease the muscles and relieve any back pain.

common causes of back pain:

- poor working posture and inadequate or infrequent exercise
- extended periods of working in a fixed position
- poor car seat design
- lack of exercise, sagging mattresses, and badly designed home seating
- unaccustomed manual work
- the later stages of pregnancy

WHEN TO SEEK PROFESSIONAL ADVICE

Even though back pain is very common there are times when its management may require the professional guidance of an osteopath, chiropractor, or a doctor. To help you decide if you should seek professional help, read this section carefully before starting to treat back pain at home.

Seek professional advice if
- you have pain in the leg that is worse than the back pain itself
- you have pins and needles or pain that radiates to the foot and/or toes
- you have a numb leg and/or foot
- your back pain is getting progressively worse
- you have a previous history of cancer
- you have had a recent infection
- you have a fever over 100°F (38°C)
- you have used steroid medication for a long time
- your back pain is worse after rest
- you have unexplained weight loss
- you are losing muscle strength in your legs.

shiatsu stretching for the shoulders

Shiatsu techniques for stretching the shoulders can be beneficial in easing neck pain, especially where tension is felt across the tops of the shoulders. Stretching the shoulders will also help the muscles surrounding the shoulder blade and mid-back.

1. With your partner lying on his or her back, grasp the forearms and give a slight pull upward and gently backward to stretch the shoulder muscles.

2. With your partner lying on his or her side, hold the shoulder cupped firmly in both hands. Rotate it comfortably (do not force it). After a few rotations lean back allowing your weight to stretch the shoulder and neck.

shiatsu stretching for the hips

Stretching the hip muscles will generally improve the flexibility of the hip and leg. It is worth noting that the hip muscles also provide support for the back, so inflexibility in the hips can often be the cause of back tension leading to back pain. There are many effective shiatsu techniques to help loosen the hip muscles.

1. Raise both knees at right angles to the body. Gently and fluidly ease the knees toward the chest, asking your partner to tell you if the move starts pulling or hurting, which should be avoided.

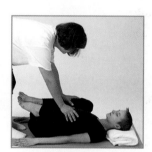

2. Once a gentle stretch has been achieved, slowly and fluidly move the knees from left to right a few times.

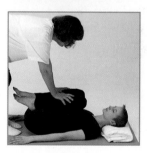

ARMS, HANDS, LEGS, AND FEET

The limbs often suffer from overuse injuries, many of them occupational. Many typists, for example, suffer from painful fingers and wrists as a consequence of word processing. After some time, the body accommodates around this and the pain becomes unnoticeable. However, many typists may go on to develop pain and stiffness in their forearms and, after many years, even suffer damage to their joints.

Overuse injuries often occur in sports. Tennis elbow, for example, in players who have poor technique causes strain to the muscles in the forearm and creates an inflamed elbow joint.

Sprains and strains of the leg and arm joints and muscles are also very common in people who play sports. Touch therapies can be invaluable in helping to alleviate pain and stiffness.

Acupressure

Applying pressure to key acupoints on the body can help relieve aches and pains in the arms, hands, legs, and feet.

SI9

SMALL INTESTINE (SI)9: Found just below the shoulder joint at the end of the crease of the arm, this point helps to increase the mobility of the arm and shoulder. It can also be used to reduce swelling and pain. Once located, apply finger pressure in an upward direction toward the shoulder blade.

LARGE INTESTINE (LI)10: This is an important acupoint for the treatment of tennis elbow and other forearm muscle problems and elbow pains. It is found just below the crease of the elbow in the central belly of the muscles that allow the backward bend of the wrist joint. Use gentle thumb pressure over this point, directing the force toward the bones of the forearm.

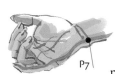
LI10

PERICARDIUM (P)7: Used traditionally as a point to help stimulate the blood circulation and to prevent faintness, this powerful acupoint can help to relieve carpal tunnel syndrome and wrist joint arthritis. It is easily located in the center of the wrist crease, in line with the middle finger. Make sure you use gentle thumb pressure over this highly sensitive point.

P7

GALLBLADDER (GB)30: This is a classic acupoint in Traditional Chinese Medicine for the treatment of hip joint arthritis and sciatica. It can be found two-thirds of the way from the sacrum (tail bone) to the tip of the hip in a depression just under the thigh bone. Press firmly with a knuckle into this point, directing your force of your pressure to the front of the hip joint.

GB3

SP10

SPLEEN (SP)10: By applying a firm fingertip force to this point, found on the inner part of the knee, relief can be gained from arthritic knee pain as well as from simple strains. In TCM (Traditional Chinese Medicine) this point also helps stimulate circulation and improve skin conditions.

URINARY BLADDER (UB)62:  UB 62
This acupoint is found on the outer side of the ankle joint directly below the tip of the ankle joint. It is a sensitive point to stimulate, but firm fingertip pressure can help in cases of aching legs, lower back, and ankle areas. Apply the pressure slightly down toward the little toe.

Aromatherapy massage

Regular massage can help to stretch muscles, making injury less likely.

As the body ages, the muscles lose a certain amount of elasticity and people become less supple. This can place strain on the joints and can lead to wear and tear injuries. Much of this can be prevented by regular massage and gentle exercise.

FOR ARTHRITIS: Mix together the following essential oils and use in a massage treatment:

- 8 drops birch
- 6 drops lavender
- 6 drops marjoram
- 4 drops ginger
- 1.5 fl.oz. (50ml) carrier oil, such as almond oil

Reflexology for sciatica

Pressure on the sciatic nerve due to a lower back problem can cause both pain and disability. Simple reflexology can help ease the symptoms by encouraging the body to release natural pain killers, known as endorphins. Work along the foot following the sciatic nerve area on the diagram on page 41, repeating the thumb stimulation about ten times; use this touch therapy three or four times each day for effective pain relief.

leg massage

Try this soothing leg massage to ease tired leg muscles and joints. Ask your subject to lie on his or her front and ensure that your subject is comfortable (use supporting towels as necessary).

1. Gently slide both hands from above the ankle, lightly over the back of the knee to the top of the thigh. Glide both hands over the hip and slide back down to your starting point. Repeat a few times.

2. Next, using your thumbs, gently stroke up the calf using small circular movements. Repeat a few times and then effleurage from the ankle to the front of the knee.

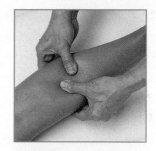

3. Gently pick up the calf muscle with one hand and pass it to the other. Repeat a few times before stroking from the ankle to the knee. Make circular movements with your palms up the front of the thigh. Repeat a few times then repeat the sequence on the other leg.

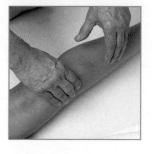

Skin

The skin is the largest organ of the whole body. As well as covering the entire body and protecting the internal organs, the skin also has an important role to play in regulating body temperature. It is also one of the body's major excretory organs for the elimination of waste products through sweat. The skin contains many sensory nerves, which vary in concentration throughout the body, the most being found in the fingertips.

The skin reflects the state of our health. When we are stressed or are eating a poor diet, the skin can become dry or oily. A clear complexion is a good indication that our general health is good. Being a main excretory organ, waste products are lost through the skin, so a poor diet can result in spots and oily skin. On the other hand, lack of fluid or excessively hot or humid conditions can produce dry skin. As age increases, the skin loses suppleness and elasticity.

The skin consists of two main layers, the epidermis and the dermis, plus a subcutaneous fat layer. Each layer has a specific function to perform in maintaining skin health and vitality. The outer layer of skin, the epidermis, is a flexible waterproof layer that protects us against the environment. Under this is found the dermis, which contains many blood vessels and specialized nerve endings and skin glands vital to a balanced complexion.

At the deepest layer the subcutaneous fat anchors the skin to the body and forms the cushioning and body contours that make our physical appearance so individual and unique.

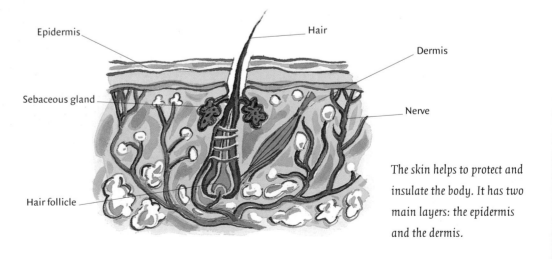

Epidermis

Hair

Dermis

Sebaceous gland

Nerve

Hair follicle

The skin helps to protect and insulate the body. It has two main layers: the epidermis and the dermis.

THERAPIES FOR GLOWING SKIN

The skin is exposed to so many environmental factors, such as pollution, wind, and sun, that a general cleansing treatment is recommended for all types – oily, combination, or dry.

Try a simple steam treatment, using a few drops of lavender, rosemary, and geranium oil as described in Chapter 2, page 33. Follow with a relaxing face massage, using a few drops of lemongrass mixed with grapeseed oil (see right).

While a regular facial massage is an easy way to help blood flow to the skin, there are other simple methods that can be equally helpful.

Many of the therapies involving treatment via acupoints can be applied quickly to promote healthy skin. Acupressure to Large Intestine 4 and Spleen 10, for instance, can help relieve skin irritations by regulating the general circulation.

Excess waste products (toxins) in the body often cause poor skin complexion. Applying thumb-walking reflexology techniques to each foot will help the excretion of unwanted toxins.

AVOIDING SUNBURN

The following essential oils can increase the risk of sunburn. These oils are said to have a photosensitizing effect and should be avoided if you have sensitive skin, are undergoing sunbed treatment, or if you are exposed to strong sunlight.

- angelica
- bitter orange
- lime
- cumin
- lemon
- orange

relaxing facial massage

Sit comfortably and wrap long hair in a towel or pin it back from your face. Apply the oil mixture to your hands and rub them together, allowing the oil to warm up a little.

1 Start by gently effleuraging the face from the chin, across the cheeks and temples, and over the forehead. Repeat the stroke four to five times or until your face starts to relax.

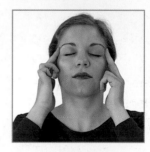

2 Next, use soft, circular, finger kneading movements across the chin, cheeks, temples, and forehead.

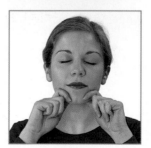

3 End by lightly applying pétrissage to the chin, cheeks, and forehead areas. Finally, relax the entire face again with effleurage.

Allow about 30 minutes to complete this home facial treatment. Afterward your skin will feel vibrant and clear, and your circulation will be boosted. End by running your fingers through your hair and applying pressure to stimulate your scalp. This will stimulate nutrition to the roots of your hair. If you wear contact lenses, remove them before you start.

essential oils for beauty treatment

For normal skin, add one of the following oils to 1 tablespoon (15ml) almond, hazelnut, or apricot kernel carrier oil:

- geranium
- palma rosa
- jasmine
- bois de rose
- frankincense

1. Start with a gentle stroking on the neck. Massage from left to right in a rhythmical upward movement. Finish with gentle fingertip "pinching" of the front and sides of the neck for a few minutes to boost blood circulation.

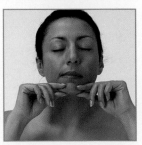

2. With the first two fingers of each hand, work along the undersides of the cheekbones, gently pressing into the tissues. Work toward the jaw, repeating five times or more. Work areas of congested tissue a little more.

3. Place your thumb pads under the lower part of the jawbone and support with the first two fingers held over the bone. Use pinching movements to work back and forth along the jawline toward the chin. Repeat five times.

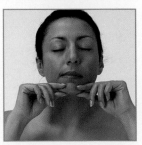

4. Place your index finger pads on each side of the bridge of your nose. Work down toward the nostrils using circular massage directed inward toward the opposite finger tip. Repeat for about 2 minutes.

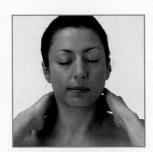

5. Using a gentle, fingertip, circular massage start at the outer edge of each eye socket and work toward the nasal corner, first along the upper and then lower border of each. Work both sockets together and repeat ten times.

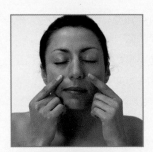

6. Place your thumbs at the outer edges of your eyebrows and your index fingers at the inner edges. Gently pinch between the two to stimulate the circulation and muscle tone. Repeat after about 2 minutes.

THERAPIES FOR SKIN PROBLEMS

There are many types of skin problems from eczema, which tends to be most common on the hands, ears, feet, and legs, to acne, which is suffered to some degree by more than 70 percent of teenagers.

Aromatherapy massage

For problem skins, try following essential oil combinations in the facial massage shown on page 97.

OILY SKIN TYPES:

- basil
- sage
- ylang-ylang
- eucalyptus

DRY SKIN TYPES:

- sandalwood
- peppermint
- geranium
- lavender

SKINS PRONE TO SPOTS:

- tea tree
- clary sage
- lavender
- thyme

Acupressure

When a whole body approach is needed for more systemic skin problems, such as eczema or psoriasis, acupressure treatment can be of great benefit.

LI4

LARGE INTESTINE (LI)4: When the skin is very irritated, use the point known as Large Intestine 4, found on the web space between the thumb and first finger. This point has an important place in Traditional Chinese Medicine because of its far-reaching effects on the whole body. Stimulate it with finger pressure every hour for 2–3 minutes until the irritation subsides.

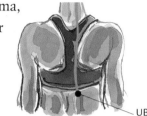
UB18

URINARY BLADDER (UB)18: For skin inflamed with psoriasis and eczema, use the Urinary Bladder point 18. This has detoxifying powers and is used in conditions that require blood purifying. Psoriasis and eczema are considered to be toxic, systemic conditions requiring a general detoxification of the body in order to improve. Locate the point two fingerwidths away from the spine, just above the waistline.

Reflexology

Stimulate the following zones for a few seconds. Refer to the reflexology chart on page 41 for more information.

THYROID REGION: use "thumb walking" to regulate skin metabolism.

PITUITARY REGION: use "thumb walking" to regulate hormone levels.

KIDNEY REGION: use thumb pressure to facilitate elimination of toxins.

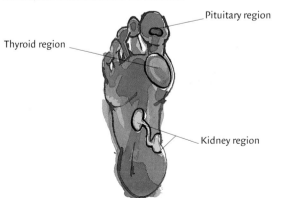
Pituitary region
Thyroid region
Kidney region

The nervous system

The nervous system is made up of the central and peripheral nervous systems. The brain and spinal cord, which form the central system, act a bit like a computer by controlling many millions of signals and inputs every second. The peripheral system can be thought of as functioning like cables and wires that connect all the body tissues and organs to the central system, which, in turn, processes the information received from the periphery.

A functional division can be made between the different nerves that comprise the peripheral system. Some function on an automatic level (the autonomic system) and help fine tune the body systems, such as circulation; sensory nerves detect changes in the environment, and motor nerves make muscles contract to help move limbs and joints. The autonomic system can be further divided into nerves that speed up the body, known as sympathetic nerves, and those that calm down the body, known as parasympathetic nerves.

For mental well-being, the nervous system must be balanced or symptoms, such as fatigue, tiredness, nausea, constipation, diarrhea, and insomnia, can ensue. Imbalances are usually due to pressure, and touch therapy can help to maintain a state of relaxation and alleviate the symptoms associated with stress.

The fight and flight reaction
Stress is a common factor or even the founda-tion of ill-health. Our bodies, however, have a

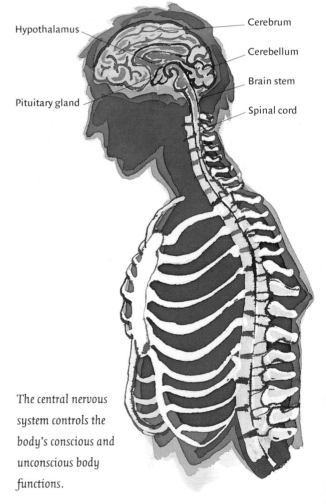

Hypothalamus

Pituitary gland

Cerebrum

Cerebellum

Brain stem

Spinal cord

The central nervous system controls the body's conscious and unconscious body functions.

limited number of ways to deal with it. Thousands of years ago stress would have been a short-lived experience involving an attack from a wild animal or other aggressor. Our bodies would have jumped into action, thanks to the sympathetic nervous system, stimulating an increased heart and breathing rate. If all went well, the experience of stress would have passed rapidly as we won the fight or escaped to safety. However, in modern life stresses commonly last for days, months, or sometimes years, but our nervous reactions are the same as they were thousands of years ago, forming the basis of stress-related illness.

RELAXING THE NERVOUS SYSTEM

Holistic treatments, either carried out at home or by a specially trained practitioner, can have profound effects on the functioning of the nervous system. You just have to start a simple aromatherapy massage treatment to feel how your body starts to relax and release stored negative energy and muscular tension. You can also improve your emotional outlook on life.

Healing

Because good health depends on the free flow of energy, energy blockages or imbalances from physical causes, injury, emotional stress, drugs, and grief can all cause ill-health. Healing can help channel positive energy in a focused way into the body, unblocking any congested energy pathways and releasing the emotions. Many people suffering from shock, unresolved grief, depression, or negative thinking can be helped enormously by healing.

Reflexology

There are many ways to use reflexology for relaxation and there is no set formula for doing so. Feel free to try any combination of the strokes shown on page 100, but bear in mind the golden rule for relaxation: keep strokes SLOW and SMOOTH.

Acupressure

Pressure applied to key acupoints will aim both to rebalance energy flows around the body and calm the nervous system.

GOVERNOR VESSEL (GV)20: This point is found on the top of the head. Ask a friend to apply a firm, but effective pressure over this point for about 5 minutes.

URINARY BLADDER (UB)2: These points are easily reached by placing your finger and thumb on each side of your nose at eye level. Stimulate these points simultaneously for the best results using gentle finger pressure for 2–3 minutes.

GV20

UB 2

Massage

Try the relaxing back massage on page 102 to help calm the nervous system. Massage is an extremely effective touch therapy for nervous conditions since it acts on a physical level, relaxing muscles and relieving tension, and on an emotional level, calming and soothing away stresses and strains that are often at the root of nervousness. Stimulating one part of the body can help alleviate problems elsewhere.

reflexology to relax the nervous system

Try to devote about a half hour to this reflexology treatment and don't rush it! Refer to the reflexology section on page 41 for a comprehensive reflexology map of the foot reflex zones. Explain the concepts of abdominal breathing and encourage your subject to practice the technique during the treatment. This will help your subject to relax as well as rebalance the energy flow in his or her body.

① Squeeze the foot together to find the hollow that locates the solar plexus.

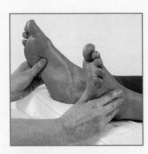

② Then relax each foot with thumb pressure over the solar plexus reflex zone.

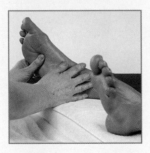

③ Using thumb pressure, "walk" up over the following reflex zones: heart, thyroid, spine, pancreas, and liver, first on the right foot then on the left foot.

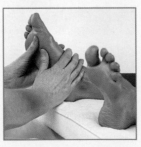

④ Locate the following reflex zones and apply thumb pressure, using a gentle rotating motion to stimulate the following points: lungs and chest, adrenals, pituitary, and kidney.

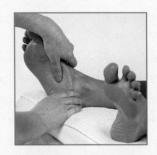

⑤ Apply gentle pressure over the solar plexus again on each foot.

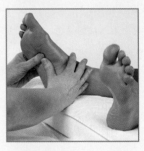

⑥ Finally, lightly run your fingers up and down each foot from the ankle to the toes and back again.

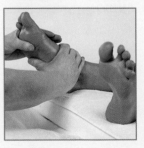

RELIEVING STRESS AND FATIGUE

First you need to identify the cause of your anxiety and stress. This is essential to understanding your fears and making the best decisions on how to overcome them. It is surprising how many people feel stressed and anxious but cannot put their finger on the cause. There is no simple, one-step method to conquer anxiety or stress, but the following therapies may form an effective home-management strategy of this common problem.

Aromatherapy massage

Mix an antianxiety essential oil formula (see below) and apply dabs to your wrists and neck when you feel under stress. This will help you control some of the symptoms when you are out and about. Try the relaxing neck and shoulder massage on page 89 using the oils described.

Acupressure

Learn to find the acupoints Heart 7 (*see page 117*) and Pericardium 6 (about 2 inches further up the inside wrist). Pressure applied to these points will help calm your nerves and mind, which are essential factors in effective anxiety control.

essential oils for stress reduction

Combine the following oils with 2 fl.oz. (60 ml) almond carrier oil. Keep the mixture in an airtight container and carry it with you to apply in times of anxiety or stress.

- 6 drops lavender
- 3 drops orange
- 2 drops marjoram

Reflexology

Using reflexology will not take away the root cause of your stress, but it will help you shake off some of its detrimental side effects.

Treatment using the reflex zones suggested below will help promote better general well-being and make you less irritable, while the more far-reaching effects of stress, such as poor immunity can be rebalanced and brought back to health. See pages 40–41 for more detail.

Catching stress in its early stages helps avert long-term illness. Stress causes a number of debilitating conditions, especially those related to the digestive and respiratory systems.

KEY REFLEXOLOGY POINTS FOR STRESS

Zone	Effect on body
Adrenal	Reduces stress and fatigue by rebalancing adrenal function
Pituitary	Regulates glandular function, reduces fatigue
Lungs and chest	Improves breathing, helps reduce hyperventilation
Thyroid	Balances metabolic rate
Heart	Calms the heart, promotes balanced circulation
Spine	Reduces nerve and muscle tension
Pancreas	Steadies blood-sugar metabolism
Kidney	Regulates fluid balance
Solar plexus	Calms and relaxes the entire nervous system

Carry out this relaxing back massage on a stressed partner or friend to help calm the nerves. Ensure your subject is comfortable and the room is well prepared. Warm the essential oil mixture (right) in your hands before you start. Finish off the massage by using long, sweeping, effleurage movements from the mid-spine up.

essential oils for calming nerves

Mix together the following oils in 100g almond carrier oil for a relaxing and soothing massage oil:

- 8 drops lavender
- 4 drops bergamot
- 2 drops sandalwood
- 2 drops camomile
- 3 drops ylang-ylang

1 With your thumbs on each side of the spine and using light circular movements, work to the upper back from the spine base. Lightly glide down the sides of the body with both hands and repeat the circling up the spine.

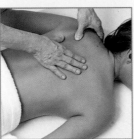

2 Repeat the same movements, this time working your way up from the lower back to the mid-back, gently easing and soothing areas of built-up tension.

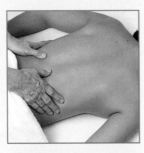

3 Pick up each large shoulder muscle (trapezius) with one hand and gently squeeze it. Pass it to the other hand and gently squeeze. Work the muscle, starting in the midline and working out toward the shoulder.

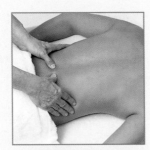

4 Then pick up the muscle again with both hands as before, but this time slide your thumbs across the muscle fibers. Repeat on the other side before using the long effleurage movements to complete.

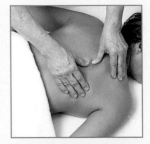

5 Place your thumbs just under the base of the scapula (shoulder blade). Slide each thumb a short distance along the scapula edge, pressing slightly. Repeat a few times and move to the other side.

6 Pick up the muscles around the waist with one hand and pass to the other hand. Work toward the hips and then back again. Repeat on the other side. Pick up the muscles and slide your thumbs across the fibers, but don't pinch the skin.

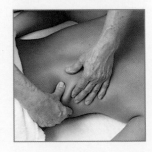

IMPROVING MENTAL WELL-BEING

Depression can be a serious psychiatric condition, and moderate to severe depression will require professional help. However, simply "feeling low" can be overcome, and many of us can manage this effectively at home, without resorting to medical treatments, using a combination of simple touch therapy treatments.

Aromatherapy massage

By using a combination of the following oils in conjunction with massage, the spirits can be lifted and well-being restored.

- ylang-ylang
- lavender
- rose
- bergamot
- neroli
- geranium
- lemon

Reflexology

Turn to the reflex zone map on page 41 and apply a stimulating thumb pressure treatment to the following points to help improve a partner's mental well-being:

- solar plexus
- heart
- thyroid
- brain
- adrenals
- lungs

spirit-lifting massage

essential oils for lifting the spirit

Mix the following formula in 8 fl.oz. (250 ml) almond carrier oil to use in the step-by-step spirit-lifting massage below:

- 24 drops bergamot
- 12 drops geranium
- 2 drops neroli
- 6 drops rose

① Apply slow, rhythmic effleurage to the back and arms (and to the legs if you wish). Keep your fingers together, following the contours of the body and use long strokes that work the oils into the skin.

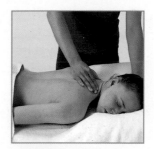

② Where you find areas of muscular tension, knead the skin gently to "iron" out muscle tissue. Work systematically across tense areas, rolling away any knots.

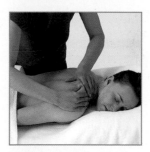

③ Keep your hands relaxed, but hold your fingers together to apply a warming circular massage movement to stimulate the skin and muscles. Keep your hands moving rhythmically for best results.

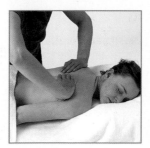

Respiration

This system includes the nose, lungs and related airways, ribs and surrounding muscles, and the diaphragm. Together, they enable us to breathe in oxygen from the air and exhale carbon dioxide, a waste product produced by cells. Our breathing rate alters dramatically when we are under pressure and it is possible to control and slow our breathing. This helps us avoid the problems of rapid breathing or hyperventilation when the concentration of blood chemicals in the brain alter to produce symptoms of dizziness and fainting. Massage and aromatherapy can also help to keep us calm and slow our breathing rate.

In holistic terms the respiratory system forms a vital connection between the inner energy needed for life and a healthy mechanism for excreting unwanted waste products and toxins. Traditional Chinese Medicine places great emphasis on lung function, relating many conditions, such as allergies, lethargy, and poor circulation, to disordered lung energy. Treatments aimed at improving lung function focus on those acupoints that have a toning and strengthening effect on lung chi.

Contained within the chest cavity, with the lungs and heart, is a collection of lymph tissue and vessels. The lungs and respiratory system are often the first tissues to come in contact with invading bugs. Maintaining the ability to ventilate the lungs with fresh air and optimize the available capacity for oxygen exchange promotes an improved resistance to infection and illness. The function of the lungs is therefore essential in our ability to fight infection and disease.

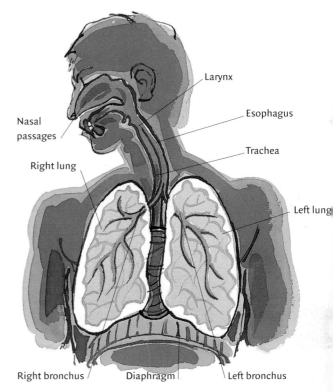

Larynx

Esophagus

Nasal passages

Trachea

Right lung

Left lung

Right bronchus   Diaphragm   Left bronchus

The body's respiratory system allows oxygen to be breathed in, which is essential for normal body functions.

COMMON RESPIRATORY PROBLEMS

As pollution increases, the number of breathing problems also rises. Children are suffering more than ever from respiratory diseases such as asthma. The following treatments can help ease the symptoms of many common breathing problems and diseases.

Acupressure

Applying pressure to key acupoints will help relieve irritating symptoms such as excess mucus and congestion.

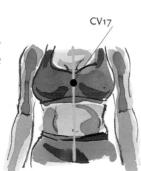

CONCEPTION VESSEL (CV)17: For a powerful, stimulating effect for the lungs, apply a circular pressure over Conception Vessel 17, located in the middle of the chest (over the breast bone) between the nipples.

GOVERNOR VESSEL (GV)14: As well as helping with a stiff neck, this strong point can help to fight the common cold. Apply strong pressure over this point, which is located at the inner tip of the shoulder blade, for 2–3 minutes.

STOMACH (ST)40: This point lies on the outer part of the lower leg, halfway between the ankle and the knee cap. This distal point can be easily stimulated using finger pressure for 2–3 minutes and will help to release congested mucus.

Aromatherapy massage

Try the following essential oils either for a simple steam inhalation or as oil for a therapeutic massage.

COUGHS AND COLDS:
- bergamot
- eucalyptus
- lavender
- marjoram
- peppermint
- rosemary

CHEST INFECTION:
- eucalyptus
- peppermint
- sandalwood
- lavender
- frankincense

CATARRH AND SINUSITIS:
- eucalyptus
- lavender
- peppermint
- rosemary
- tea tree

Reflexology

Stimulate the following key zones for a few seconds each:

CHEST REGION: "walk" the thumbs here to help relieve congestion.

BRONCHIAL REGION: use thumb "walking" here in order to help relax asthmatic lungs.

DIAPHRAGM REGION: use thumb pressure to facilitate breathing.

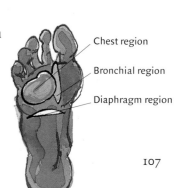

Chest region

Bronchial region

Diaphragm region

massage to release congested mucus

Find a low table so that your partner can lie face down with the table supporting the lower body. The upper body should hang over the edge of the table with the head resting on a pillow. Alternatively, find at least three plump pillows and, again with your partner lying face down, place the pillows underneath the stomach. Place a bowl nearby for your partner.

1. Using an appropriate oil, see page 107, and long, sweeping effleurage strokes, move from the waist up toward the neck at least eight times.

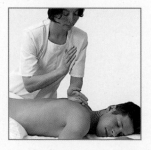

2. Cup your hands and apply one hand to the mid-back followed by the other. Repeat this at a rhythmic pace all around the mid- to upper-back region. Effleurage the area. Ask your subject to cough mucus into the bowl.

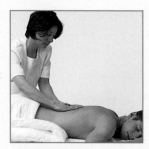

3. Then, with the edge of your hands, apply one hand and then the other, keeping the wrists loose, to the same area of the back. Repeat the effleurage again. Repeat this whole process a few times.

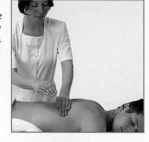

acupressure to help the sinuses

Severe congestion and pain around the upper face and forehead can develop when the passages that lead from the sinuses become blocked with mucus. This can follow a cold and, if left untreated, sinusitis can develop, in which case you should visit your doctor. Acupressure, however, can help relieve the symptoms.

1. Find acupoint BL10, which is located on each side of your spine about two fingerwidths below the base of your skull. Apply gentle pressure here for 10 seconds.

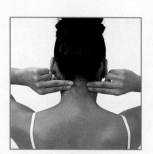

2. Next, locate acupoint K27, which is just under your collarbone on each side of your breastbone. Apply gentle pressure here for 10 seconds.

3. Finally, press your thumb knuckles on each side of your nose in the area of your sinuses. Work out along the cheekbones applying gentle pressure as you go, but not causing pain.

Shiatsu

Breathing is essential for life and therefore vital to a healthy circulation of ki (energy). Shiatsu methods of optimizing the breathing look toward the Hara, or abdominal cavity, as the key to health and success. Try this simple breathing exercise to improve your breathing.

Lie on the ground or sit with your spine straight. Tense all the muscles in your body as you inhale. Concentrate on the feeling of tension you have generated and hold it for about 10 seconds.

Gradually release the tension and relax your muscles as you breathe out and try to let go completely. Repeat this three to four times.

To ensure you have mastered abdominal breathing, inhale very slowly through your nose while pushing out your abdomen and moving your chest as little as possible. After your abdomen is stretched, expand your chest with air. Hold this breath for about 5 seconds, then slowly breathe out. As you do, relax your abdomen. Repeat this method 15–20 times and focus your attention on the movements of your abdomen as you breathe in and out. After a while you will automatically breathe using your lower abdomen.

GIVING UP SMOKING

Try using the following acupressure points when you feel the desire to smoke:

LUNG (LU)6: Firm pressure over this point helps open up the chest and clear the lungs of toxins. Lung 6 is found about halfway between the wrist and elbow in line with the thumb.

LU 6

EAR ADRENAL ACUPOINT: Acupuncturists insert a semi-permanent stud into this point, which allows it to be stimulated when needed. It can just as easily be self-stimulated using a pointer, such as a blunted toothpick or cocktail stick. Once stimulated, this point increases the adrenal function and will help to calm and reduce cravings for nicotine.

Ear adrenal acupoint

---

good breathing is needed for

- healthy exchange of blood gases and cell nutrition
- better management of stress
- better lymph circulation and drainage
- improved feeling of general well-being

THE EFFECTS OF BALANCED BREATHING

Natural and effective breathing is not attained simply by increasing your oxygen intake. True balanced breathing is only achieved by optimizing the distribution of oxygen throughout the entire body. Through focused meditation you can drop your respiration rate to 6 breaths per minute, thus reducing your oxygen intake by 10%. This is balanced by a 300% increase in the circulation of blood through the muscles. Meditation and breathing techniques such as the Hara method have been shown to exert their effects by redistributing blood and oxygen.

Digestion

Without an adequate digestive process our bodies could not remain healthy for long. Most of the nutrients required for life are derived from our food, which must be broken down by the enzyme action of the digestive juices released by the stomach, intestines, and pancreas. However, digestion starts in the mouth with the action of the enzyme amylase, which is present in saliva.

Chewing is a vital aspect of digestion; it not only breaks down food into smaller pieces but also mixes the food with enzymes produced in the mouth. Once activated by enzymes the rest of the digestive process follows in a well-orchestrated sequence. It is not uncommon, however, for this process to be upset by a number of factors that results in the digestive process being thrown out of balance.

Stress is among the most powerful disrupters of digestive function. The body's response to stress causes the release of adrenaline, which causes the stomach and gut to slow down. A sluggish system does not efficiently process food, and discomfort, bloating, and altered bowel function occurs. This can cause a variety of problems ranging from heartburn, stomach cramps, irritable bowel syndrome (alternating constipation and diarrhea), and ulcers.

Our mental state also tends to influence what we eat – many people will consume too much or too little food, or eat the wrong types of foods when feeling low. Touch therapies can help improve general digestive health.

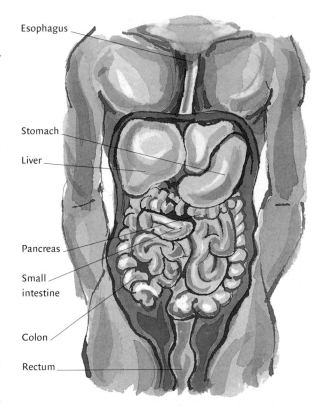

The digestive system starts in the mouth and includes the esophagus, stomach, small intestine, and large intestine; it ends with the elimination of waste at the anus.

IMPROVING DIGESTIVE HEALTH

The use of touch therapies for digestive problems can be of great benefit for sufferers of common problems such as irritable bowel syndrome and constipation. However, any noticeable change in your bowel movements should first be reported to your doctor for a proper diagnosis.

You should not treat anyone with a touch-based treatment for digestive problems soon after they have eaten a meal. And, before attempting to self-treat any digestive condition, take some time to familiarize yourself with your abdomen, using the simple self-massage procedure outlined to the right.

Acupressure

Apply pressure to these key acupoints to help stimulate and balance the digestive system.

STOMACH (ST)25: Find this point by measuring three fingerwidths away from your navel on either side. Apply firm pressure directly downward into the abdomen for no longer than 2 minutes.

Stomach 25 is traditionally used for stimulating the large bowel. This point has powerful effects on the digestive system and should never be overstimulated.

CONCEPTION VESSEL (CV)12: To help regulate appetite and balance digestion, locate this point halfway between your navel and the lower edge of your breast bone.

Using gentle finger pressure and apply stimulation for 2–3 minutes.

abdominal self-examination

It is important you become aware of how your abdomen feels under normal conditions before you will be able to identify any areas of congestion or discomfort. Get used to this self-examination before you embark on any form of self-treatment.

1 Find a comfortable surface to lie down on and settle yourself using pillows under your knees and head.

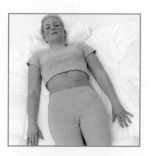

2 Hold your fingers together lightly and gently press your abdomen up and around under the ribs, finishing low down left of your abdomen. You may find areas of tenderness or congestion.

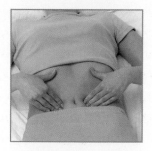

3 Continue to dip your fingers in a circular fashion working inward toward your navel. Any discomfort felt during the first exploration could suggest a problem with the large bowel, whereas any felt during the second exploration may suggest the small intestine.

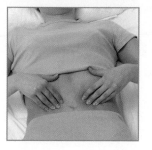

Reflexology

Stimulate or relax the digestive organs, as required, by following the chart below. Keep in mind that the more vigorous the stimulation, the greater the reaction. Always keep treatments calm and slow if the organs require relaxation and only stimulate for sluggish or under-functioning digestive organs.

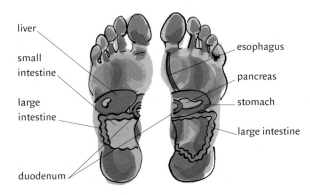

liver

small intestine

large intestine

duodenum

esophagus

pancreas

stomach

large intestine

RELIEVING CONSTIPATION

A common digestive malfunction in the West, constipation is often a result of eating too many highly refined foods. It can be caused by stress and disease, though, so always visit your doctor for a diagnosis if constipation persists for more than a couple of days.

Aromatherapy massage

Try using the essential oil blend described opposite in combination with the step-by-step massage strokes outlined opposite to help bring relief to constipated bowels.

Other treatments for constipation

Although dietary imbalances are often cited as being the main cause of constipation, many other factors come into play. Stress and anxiety can rapidly alter bowel function just as easily as a lack of dietary fiber. Too much fiber in the diet can sometimes have a constipating effect. For the bowels to work healthily and effectively you should also drink at least 3–4 pints (1.5–2l) water daily.

Shiatsu for digestive problems

Shiatsu can help ensure that the digestive system is working at its optimal level. A practitioner can diagnose and offer effective help for a wide range of conditions, such as:
• diarrhea – working on the spleen and kidney meridians
• constipation – working on the large intestine, stomach and gall bladder meridians
• indigestion – working on the liver meridian
• general stomachache – working on the stomach and spleen meridians.

Shiatsu massage is very safe and highly effective, but we recommend consulting a qualified practitioner. Abdominal treatment is not really suited for home use.

WHEN TO SEEK PROFESSIONAL ADVICE

Most cases of constipation are not associated with any cause for concern. However, if bowels are normally regular, seek advice if a bowel motion is not passed in 48 hours or there is pain and/or bleeding when you attempt a bowel movement.

REFLEXOLOGY FOR CONSTIPATION

By stimulating special reflex zones on the sole of the foot, reflexology can help promote a balanced, healthy bowel function. Regular treatment will ease constipation and effectively eliminate toxins from the body.

Start by working over the iliocecal valve zone (illustration 1). Use sustained thumb pressure over this point for about 10 seconds. Next, follow the course of the ascending and transverse colon zone, working the thumb across the sole of the foot toward the outer edge of the foot. Finish off by stimulating the descending and transverse colon in a similar fashion but with thumb strokes working toward the inner side of the foot.

essential oils for indigestion

- lemon balm
- allspice
- caraway

Nausea:
- ginger
- mint
- camomile

essential oils for constipation

- lovage
- orange
- nutmeg

Infant colic:
- camomile
- dill
- hyssop

massage to relieve constipation

essential oils for abdomen

Mix the following essential oils in 4 fl.oz. (100ml) carrier oil, such as almond oil:

- 6 drops orange oil
- 4 drops ginger oil
- 4 drops fennel
- 4 drops peppermint
- 2 drops lavender

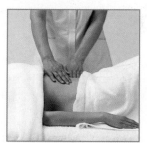

1. Ensure your subject is settled comfortably and apply the oil to your hands to warm it. Use a gentle finger dipping action to stimulate the large bowel area.

2. Apply a deep, sustained pressure, with your fingers held together, into the lower left area of the bowel or anywhere you feel any areas of congestion. This technique should never be painful.

3. Finish off by stroking the lower abdomen firmly to relax the muscles.

Circulation

A healthy heart and efficient circulation rely on good diet, adequate exercise, and reduction of factors such as stress and anxiety. Other stresses may not be so obvious. Smoking, excessive coffee, alcohol, and salt act as physiological stresses and can overwork an ailing heart.

With increasing age, circulation around the body becomes poorer. The lower limbs tend to suffer the most because they are the furthest points from the heart, especially in people who have to stand or sit for long periods of time.

Balanced diet and exercise are essential for a healthy heart and blood circulation. However, great benefit can be obtained by manual therapies that regulate blood flow and stimulate balanced nervous function that in turn helps regulate the heart. Massage has been traditionally used for this, but equally good results can be obtained from reflexology and aromatherapy.

Make a tight fist with your hand – this is about the size of your heart. During a lifetime it has been estimated that your heart beats about 3 billion times, pumping about 2,190,000 gallons of blood every year. Your entire blood volume is circulated around your body every minute and the heart pumps six times more blood around your body during exercise than when at rest. For an organ that has such a heavy workload and that has to remain in peak condition for many years, it is definitely worth taking the time and trouble to look after it properly.

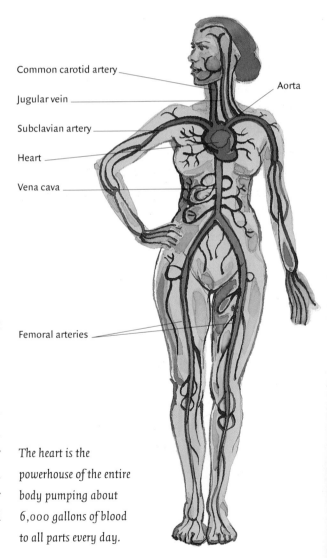

Common carotid artery

Jugular vein

Subclavian artery

Heart

Vena cava

Aorta

Femoral arteries

The heart is the powerhouse of the entire body pumping about 6,000 gallons of blood to all parts every day.

STIMULATING THE CIRCULATION

A healthy, efficient circulatory system is vital to maintain body functions. Disorders such as poor circulation can be helped by touch therapy.

Massage and aromatherapy massage

When the circulation is sluggish, blood does not always circulate around the body as efficiently as it could. For centuries, massage therapy has been used with great effectiveness to assist the blood on its way.

One of the fundamental benefits of any massage stroke is improved circulation. This comes from the pushing and pumping effect the techniques have on the veins and other circulatory vessels of the body.

Massage can be made more effective by the addition of selected essential oils to benefit the circulation. Oils such as black pepper, camphor, cinnamon, and eucalyptus can be added to a suitable carrier oil and used as described in the box right for an effective stimulating massage.

CIRCULATORY TONIC BLEND

TO PROMOTE CIRCULATION:

- 10 drops lemon
- 4 drops eucalyptus
- 3 drops black pepper
- 2 drops cinnamon

CALMING OILS FOR AN OVERSTRESSED

CIRCULATORY SYSTEM:

- lavender
- camomile
- yang-ylang
- orange

massage to improve arm circulation

Position your subject on his or her back, cover them with warm towels and make sure your subject feels relaxed. Mix the circulatory tonic blend of essential oils described below left in 4 fl.oz. (100ml) almond carrier oil and warm some in your hands before you begin.

1. Start with the right arm. While holding your subject's hand with your right hand, use your left hand to apply firm downward strokes from the wrist toward the elbow and the shoulder.

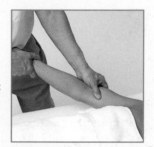

2. Follow this with a gentle kneading action along the forearm and upper arm. Do not pinch too hard.

3. Finish off by using downward effleurage strokes with your right hand, while holding the arm up at the wrist using your left hand. This helps to drain excess fluids in the arm.

massage for improving leg circulation

Legs suffer from poor circulation, especially in people who stand or sit for long periods of the day. Ensure your subject is comfortably positioned on his or her front, cover them with warm towels, and make sure your subject feels relaxed. Start by treating the uncovered right leg.

① Introduce the treatment using long, sweeping effleurage strokes, working up from the calf, across the thigh and over the buttocks.

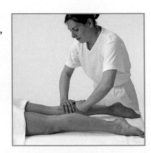

② Next, use an alternate kneading action to work the calf and thigh muscles to promote circulation.

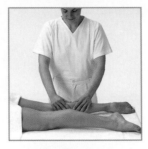

③ Finish off by using long, sweeping effleurage strokes. Next, turn your subject over and massage the front part of the thigh only, not the calf, using the techniques above.

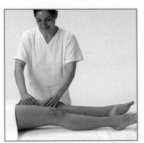

Home treatment for poor circulation and cold hands

A sluggish circulation often causes cold hands and a feeling of general lethargy and low spirits. This tends to make people more sensitive to cold weather and can aggravate preexisting conditions such as arthritis and rheumatism.

Always consult your doctor if you have high blood pressure or heart disease before undertaking any home treatment.

THE FINGER FLICK

As you draw in a deep breath, slowly make a tight fist with both hands. Keep the fists held as tightly as you can as you finish the in-breath, holding it for about 10 seconds.

Quickly snap your fists open, flicking the fingers out. As you do this, force out your breath in one exhalation. Next flap your hands and wrists loosely to flick blood to the ends of the fingers. Relax your hands and fingers to reestablish a balanced bloodflow.

THE PANTING DOG

Sit comfortably resting your hands on your knees in a slightly forward flexed posture. Regulate your breathing for about two or three in- and out-breaths.

Next increase the speed and depth of your breathing with each breath. Keep your mouth shut and breathe through your nose. Build up the deep pant to your fullest capacity, then stop and breathe gently and rhythmically again. This simple technique helps regulate oxygen and carbon dioxide levels in the blood and stimulates your heart and inner-heating energy.

Acupressure for the heart and blood circulation

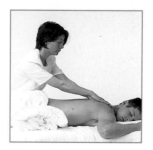

HEART 7: Find this point on the outside edge of the palm, near the crease of the wrist closest to the palm. Using sustained finger pressure, stimulate this point to improve and strengthen the heart and circulation.

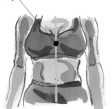

CONCEPTION VESSEL (CV)17: Located in the middle of the chest (over the breast bone) between the nipples, apply gentle pressure on this point to help ease palpitations in the chest.

KIDNEY (K)3: To improve circulation to the lower limbs, apply gentle pressure to this point, which is located on the inner side of the ankle just in front of the Achilles tendon (see right).

LARGE INTESTINE (LI)10: To improve circulation to the upper limbs, stimulate this point, which is situated on the outer side of the forearm, three fingerwidths below the elbow.

Reflexology

Stimulate or relax the heart and circulation as required by pressing the points shown on the chart below. Bear in mind that the more vigorous the stimulation, the greater the reaction. Always keep treatments calm and slow if the organs require relaxation and use a stimulating action only for sluggish cardiovascular systems.

heart area

---

massage for calming the circulation

Ensure your subject is lying on his or her front and is comfortable with pillows under the feet and abdomen. Mix one of the essential oils listed on page 113 with a carrier oil and apply a liberal amount to your hands. Rub your hands together to warm the oil before beginning the massage.

1. Start the treatment using long, slow effleurage strokes. Work from the base of the spine, up over the lower back, around the ribs and finish across the shoulders.

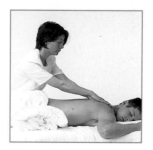

2. Keep your hands on your subject's back, but return them to the starting position by drawing them back with just the fingertips running down the spine.

3. Repeat this very simple procedure in a calm and rhythmic fashion for about 10 minutes to soothe and calm your subject's circulation.

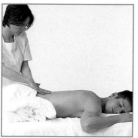

Reproduction

The health of this system is essential for the natural continuation of our species. It is this system that causes much distress if it malfunctions, especially for couples who desperately want a child. However, pregnancy can also cause anxiety for people who conceive when they had not considered having a baby, so general awareness of your system is essential.

This is one body system where the male anatomy differs greatly from the female. Whereas a woman's reproductive organs lie within the body and contain a uterus for a baby to grow, and a birth canal through which the baby travels down for birth, a man's organs, the testes and penis, lie outside the body. Women release one egg each month, while men produce and release millions of sperm from the testes with every ejaculation.

If an egg is not fertilized, it will pass down the same route that serves as the birth canal; this is known as menstruation. Many women experience problems with this monthly cycle, which are often caused by an imbalance of hormones. Symptoms range from pain, irritability, nausea, vomiting, back pain, and tiredness. Touch therapy has much to offer in alleviating many of these problems. It is important to try to regulate the production of hormones, promote circulation, and generally revitalize the whole system. Shiatsu and acupressure can prove to be very beneficial for menstrual problems, whereas regular massage is recommended for couples who are trying to conceive.

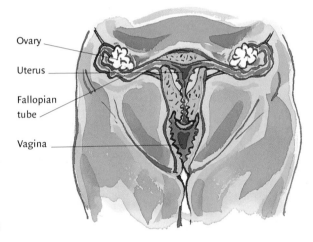

Ovary

Uterus

Fallopian tube

Vagina

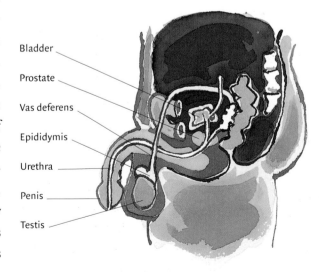

Bladder

Prostate

Vas deferens

Epididymis

Urethra

Penis

Testis

IMPROVING YOUR SEXUAL ENERGY

The health and vitality of the sexual organs features highly in Traditional Chinese Medicine. This is often misinterpreted by Westerners as a preoccupation with sexual performance, but the original emphasis was related to improving life-force. The life-force, in this case, was seen to be charged by sexual energy which then flowed throughout the whole body.

Acupressure

FOR REVERSING SEXUAL APATHY: Anxiety and tension can adversely affect sexual relationships. Acupressure can help relieve many of the negative energy blocks that inhibit a healthy relationship and pleasurable sexual intercourse.

KIDNEY (K)3: Daily stimulation of Kidney 3 can be used to improve sexual vitality.

K3

Reflexology

TREATMENT TO COMBAT SEXUAL DYSFUNCTION:

Gently rub the foot with talcum powder. Press your thumb onto the solar plexus point in the foot (see page 49). Bend the thumb at its first joint so that it moves along a little.

Use this technique over the spine and brain points (see the charts on page 41) to balance the nervous system, soothe nervous impulses, and to improve nerve flow to the sexual organs.

solar plexus

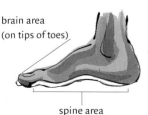

brain area
(on tips of toes)

spine area

COMBATING IMPOTENCE

At least one in four men experience erection problems at some time in their lives. Impotence, where the penis cannot become erect, can become a source of great anxiety. Similarly, premature ejaculation causes great anxiety and dissatisfaction. Touch therapy may help.

Acupressure

Applying pressure to this key acupoint can help to combat impotence.

KIDNEY (K)3: The use of sustained pressure over the acupoint Kidney 3 has traditionally been used to strengthen the genitals and can help in some cases of impotence.

K3

Reflexology

Reflexology can help dispel sexual anxieties that may contribute to impotency and loss of sexual interest. Start by stimulating the reflex zone related to the solar plexus, followed by the genital areas. A description is given on page 118.

Massage for reproductive health

The health of the reproductive organs is dependent on the circulation of blood to the pelvic region. Congestion of organs such as the uterus and ovaries may result in a fluid imbalance that could make intercourse uncomfortable. Simple abdominal massage as described on page 111 will help dispel fluid and balance the blood circulation. A combination of lavender, marjoram, and ginger oil mixed with a carrier oil such as almond oil makes an effective aromatherapy blend.

Trying to conceive can be quite stressful, especially if you have been trying for a while. However, one of the most important factors for success is that you both feel relaxed – taking time with each other is extremely important. Try this sensual reflexology exercise to help you bond with your partner and set the right mood.

essential oils for sensual massage

Combine these oils for a sensual blend that lends itself to a vitalizing massage:
- 6 drops sandalwood
- 5 drops ylang-ylang
- 2 drops vanilla
- 1 drop cinnamon
- 1 drop jasmine
- 1.5 fl.oz. (50ml) carrier oil, such as almond oil

1 Relax both feet with thumb pressure over the solar plexus reflex zone. Relax each foot separately.

2 "Walk" your thumb over the reflex zones that relate to the neck and thyroid, chest, lungs, and shoulders (see page 41). These areas relax the system and promote healthy circulation and metabolism.

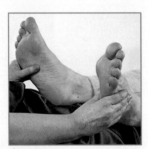

3 Use circular thumb rotation movements over the reflex zones for the uterus and prostate, ovaries, and testicles. Stimulation of these organs helps balance sex hormone production and improve sexual vitality.

4 Hold each entire foot in your hands and apply a wringing movement to the inner joints. This relaxes the foot and energizes the body.

5 Use firm, but gentle finger pressure over the reflex zone relating to the brain to improve mental attitude and well-being.

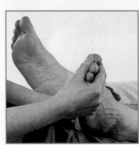

6 Finish off by using gentle massage strokes to the entire foot surface to promote circulation and balance to the system. Repeat on the other foot.

MENSTRUAL PROBLEMS

The female reproductive system is complex and sensitive to change. For its healthy and balanced functioning all aspects of life must be in harmony. Diet, exercise, and mental health are essential factors in controlling hormone balance and their regular cyclic fluctuations. Should dysfunction occur, the following home treatments may help you regain control.

Acupressure

In Traditional Chinese Medical terms, gynecological problems require stimulation of the acupoints that control vitality and longevity. These points can also be used to promote sexual health and fertility.

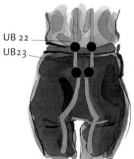

URINARY BLADDER (UB)22 AND 23: Use firm sustained pressure on the acupressure points Urinary Bladder 22 and 23 located on the inside centre of both soles and the outside of the left sole.

SPLEEN (SP)6: The distal point, located on the inside of the ankle bone – the point known as Spleen 6 – can have powerful regulatory effects on the menstrual cycle and female reproductive system.

CONCEPTION VESSEL (CV)3 AND 4: Apply firm but gentle pressure to the acupoints Conception Vessel 3 and 4. These powerful points help both to strengthen the reproductive organs and regulate menstruation.

KIDNEY (K)1: Found on the sole of the foot, Kidney 1 has a traditional role to play in the regulation of blood pressure and may help in the control of premenstrual fluid retention and menopausal hot flushes.

TRIPLE HEATER (TH)5 AND KIDNEY (K)1: Shiatsu and acupressure can help the body adjust to any hormonal changes. As with menstrual problems, massage to the lower back, sacrum, and upper back is essential. Some important points include Triple Heater 5, situated on the back of the wrist, and Kidney 1.

Aromatherapy massage

General massage to the lower back and buttocks will help give direct relief from symptoms of muscular tension and pelvic congestion.

Abdominal massage can be very effective, but must be done with great care, especially during the menstrual period itself.

Try using the essential oils of lavender or marjoram in your massage treatments for menstrual cramps.

OILS FOR MENSTRUAL WELL-BEING:
- camomile
- cinnamon
- ginger
- lavender
- melissa
- thyme

Contraindications

THERAPEUTIC TOUCH

As much as touch can be highly beneficial to many health problems, there are times when caution is needed and touch therapy should not be used. These situations are known as contraindications. The following circumstances require special caution.

Touch over the abdomen during pregnancy

Although touch is of great benefit to a pregnant woman, deep techniques applied over the abdomen could cause disruption of the normal pregnancy, especially if there has been a past history of miscarriage.

Touch on someone with cancer

This rather controversial area needs the application of common sense rather than adherence to a set of rules. There has been much talk in the massage world about the potential to move cancer cells from one area to another, previously cancer-free, site by touch because of the profound effects it has on lymphatic circulation. In reality, for patients with advanced cancers, many of whom already have secondary spread to other tissues, the positive benefits experienced far outweigh the potential risks associated with further spread. Common sense must predominate – don't use touch over the area of the tumor, avoid treatment in undiagnosed or suspected cases of cancer (or other uncertain lumps or bumps for that matter), and avoid sites that have received surgery or radiotherapy for at least 2 to 3 weeks.

Touch in cases of obvious inflammation

As a general rule, never stimulate red, hot, and swollen joints or tissues. These are signs of acute inflammation and should be treated using rest, ice, compression, and elevation (a therapy known as RICE).

Touch in a person with a high temperature

The stimulating aspects of touch can make the body tissues release more toxins, and these can have the effect of raising body temperature even further. For these reasons, therapeutic touch is not recommended until the fever has abated.

Touch in cases of disturbed circulation

One of the most common circulatory complaints is varicose veins. Touch is only really a problem in very advanced cases where the skin and dilated veins are very delicate and are liable to hemorrhage.

In other conditions, such as phlebitis and thrombosis (blood clots in veins), touch is strictly contraindicated.

Touch and skin disease

In the case of certain skin conditions, such as scabies, ringworm, and impetigo, the infection can be spread to you as well as making the condition worse for the recipient. Touch in cases where there is skin infection is, therefore, clearly contraindicated.

For other noninfectious skin problems, such as eczema and psoriasis, touch with healing oils such as lavender can be very beneficial.

AROMATHERAPY

Pure essential oils are highly concentrated and must be used with care. To use the oils safely, the following guidelines must be applied:

- do not use undiluted essential oils on the skin
- do not take essential oils internally unless advised to do so by a practitioner
- if you are pregnant or epileptic, seek the advice of a qualified practitioner
- avoid using citrus oils in the sun

Oils that can irritate the skin:

- birch
- black pepper
- cinnamon
- citronella
- clove
- cumin
- thyme
- thuja

Oils to be avoided in epilepsy

- camphor
- sweet fennel
- hyssop
- sage

Oils to be avoided for high blood pressure

- hyssop
- sage
- thyme

Caution: Never use aromatherapy during pregnancy if there have been medical difficulties or complications. Otherwise, use it only after the first three months of a problem-free pregnancy, unless you are recommended not to by an experienced health professional.

Babies and children can greatly benefit from the use of essential oils but it is important to use half the adult recommended dosage.

Potentially toxic oils to be avoided by pregnant women and children:

- hyssop
- aniseed
- basil
- camphor
- clary sage
- clove
- juniper
- majoram
- sage
- oregano
- thuja
- bitter almond
- pennyroyal
- sassafras
- savory

Resources

British Holistic Medical Association
Trust House
Royal Shrewsbury Hospital
South Shropshire
SY3 8XF
UK

British Complementary Medical
Association
39 Presbury Road
Cheltenham
Gloucestershire
GL52 2PT
UK

The General Osteopathic Council
Osteopathy House
176 Tower Bridge Road
London
SE1 3LU
UK

British Council for Chinese Martial
Arts (Qi Gong)
28 Linden Farm Drive
Countesthorpe
Leicester
LE8 5SX
UK

Register of Chinese Massage Therapy
PO Box 8739
London
UK

The General Council and Register of
Naturopaths
Frazer House
6 Netherhall Gardens
London NW3 5RR
UK

Professional Association of
Traditional Healers
190 E. 9th Avenue, Suit 290
Denver, CO 80206
USA

The British Massage Council
Greenbank House
65A Adelphi Street
Preston
PR1 7BH
UK

London College of Massage
5 Newmans Passage
London
W1P 3PF
UK

London and Counties Society of
Physiologists
330 Lytham Road
Blackpool
FY4 1DW
UK

Jan de Vries Bodywork Master Class
Auchenkyle
Southwoods
Troon
Ayrshire
Scotland
KA10 7EL
UK

Association of Systematic
Kinesiology
39 Browns Road
Surbiton
Surrey
KT5 8ST
UK

Feldenkrais Guild (UK)
PO Box 370
London
N10 3XA
UK

European Hellerwork Association
The MacIntyre Gallery
29 Crawford Street
London
W1H 1PL
UK

The British Reflexology Association
Monks Orchard
Whitbourne
Worcester
WR6 5RB
UK

Irish Reflexology Institute
4 Ruskin Park
Lisburn
Co. Antrim
BT27 5QN
Northern Ireland

The Scottish Institute of Reflexology
17 Cairnwell Avenue
Mastrick
Aberdeen
AB2 5SH
UK

The Raphael Clinic (Reiki)
211 Sumatra Road
West Hampstead
London
NW6 1PF
UK

Shiatsu Association
31 Pullman Lane
Godalming
Surrey
GU7 1XY
UK

Neals Yard Therapy Rooms (Rolfing)
Neals Yard
Covent Gardens
London
WC2
UK

Independent Professional Therapists
International
8 Oldsall Road
Retford
Notts
DN22 7PL
UK

Hadley Wood Healthcare
(Osteopathy)
28 Crescent Wood
Barnet
Herts
EN4 0EJ
UK

The National Federation of Spiritual
Healers
Old Manor Farm Studio
Church Street
Sunbury-on-Thames
Middx
TW16 6RG
UK

American Alliance of Aromatherapy
Po Box 750428
Petaluma
California 94975-0428
USA

American Aromatherapy Association
PO Box 3679
South Pasedena
California 91031
USA

National Association of Holistic
Aromatherapy
PO Box 17622
Boulder
Colorado 80308-0622
USA

Feldenkrais Guild
524 Ellisworth Street
Mount Shasta
California 96067
USA

American Massage Therapy
Association
820 Davis Street
Suit 100
Evanston
Illinois 60201-4444
USA

International Association of Infant
Massage
PO Box 438
Elma
New York 14059-0438
USA

American Shiatsu Association
PO Box 718
Jamaica Plain
Massachusetts 02130
USA

Qi Gong Academy
8103 Marlborough Avenue
Cleveland
Ohio 44129
USA

International Institute of Reflexology
PO Box 12642
Saint Petersberg
Florida 33733
USA

Centre for Reiki Training
29209 Northwestern Highway #592
Southfield
Michigan 480302-4920
USA

FURTHER READING

Fugh-Berman, A. *Alternative Medicine: What Works* (Odoninan Press, USA, 1996)

Griggs, B. *The New Green Pharmacy* (Vermillion, UK, 1997)

Harper, J. *Body Wisdom* (Thorsons, UK, 1997)

Madders, J. *Sress and Relaxation* (MacDonald Optima, UK, 1981)

Pauling, L. *How to Live Longer and Feel Better* (W.H. Freeman, USA 1986)

Pert, C. *Molecules of Emotion* (Simon & Schuster, UK, 1999)

Price, S. *Aromatherapy for Common Ailments* (Gaia Books, UK, 1991)

Tratler, R. *Better Health Through Natural Healing* (McGraw Hill, USA, 1987)

Vickers, A. *Massage and Aromatherapy* (Chapman Hall, UK, 1996)

de Vries, J. *Inner Harmony* (Mainstream Publishing, UK, 1999)

Index

Acknowledgments

The publishers wish to thank the following
for the use of pictures:
AKG 46; 52;
Bridgeman Art Library 15; 26;
Hulton Getty 30;
Imagebank 6; 61; 74
Science Photo Library 13; 14; 18; 21;
Stockmarket 25; 54
Tony Stone 11; 12; 62; 64; 66; 68; 70; 73; 76; 78; 80;
  82; 85
Wellcome Trust 42

The publishers wish to thank the following for help
with photography:
Francis Annette
C. Bayes
Susan Besley
Jan Blackbird
Richard Clarke
Jules Cloran
Ann Collins
D. Collins
Pete Donno
Sarah Douglas
Denise Hammond
Harriot Hart
Peter Hoggarth
Justin Huckle
Alice Jeavons
Amy Jeavons
Lucy Kendry
Julia Konig
Lynn McClelland

Judith Morgan
Angela Morris
Angela Neal & Family
K. Newton
Rosemary Nobbs
W. A. Ray
C. Riley
Sarah Jane Roberts
Luis Ruiz
Eleanor Scott-Plummer
F. Selkirk
Karen Simporis
Deborah Slot
C. Southern
Sundial Chiropractic Clinic
Ruby Taylor
Jawlet Turner
Pia Vuori

The publishers wish to thank the following
for help with properties
The Wilbury School, Hove, East Sussex